The Second Genesis

The Coming Control of Life

The
Second
Genesis

by Albert Rosenfeld

The
Coming
Control
of
Life

PRENTICE-HALL, INC., ENGLEWOOD CLIFFS, NEW JERSEY

*For Robert and Shana, who will be participants in, and
hopefully the beneficiaries of, The Second Genesis*

Second Printing..............May, 1969

THE SECOND GENESIS: *THE COMING CONTROL OF LIFE*
by Albert Rosenfeld

Library of Congress Catalog Card Number: 69-15329

Printed in the United States of America · T   13-797357-8

Prentice-Hall International, Inc., London
Prentice-Hall of Australia, Pty. Ltd., Sydney
Prentice-Hall of Canada, Ltd., Toronto
Prentice-Hall of India Private Ltd., New Delhi
Prentice-Hall of Japan, Inc., Tokyo

ACKNOWLEDGMENTS

In writing this book, I have been constantly aware of an indebtedness, impossible to measure, to all those scientists and physicians I have talked with and listened to and learned from over the years in the course of covering science and medicine for *Life* Magazine; and of a special indebtedness to *Life* for permission to use material gathered for its pages by myself and others.

Also, acknowledgment is hereby made for quotations of material from the following sources:

Man Adapting by Rene Dubos. Yale University Press.

Male and Female by Margaret Mead; © 1949, 1967 by Margaret Mead. William Morrow & Co., Inc.

Mechanical Man: The Physical Basis of Intelligent Life by Dean E. Wooldridge; © 1968 by McGraw-Hill Book Company. Used by permission of McGraw-Hill Book Company.

The Sanctity of Life and the Criminal Law by Glenville Williams. Alfred A. Knopf, Inc.

Sex, Science and Society by Sir Alan Parkes. Oriel Press Ltd., Newcastle upon Tyne, 1966.

The Short Novels of Dostoevsky (excerpt from the Thomas Mann introduction). The Dial Press, Inc.

Yellow Jack by Sidney Howard; *Arrowsmith* by Sinclair Lewis. Harcourt, Brace & World, Inc.

The Rebuilt Man by Fred Warshofsky. Thomas Y. Crowell Company.

Man and His Future by Gordon Wolstenholme. Little, Brown and Company.

Back To Methuselah by Bernard Shaw. By permission of The Public Trustee and The Society of Authors, London.

The Nature of Human Nature by Alex Comfort; *Profiles of the Future* by Arthur Clarke; *Science and Human Values* by Dr. Jacob Bronowski. Harper & Row, Publishers, Inc.

The Prospect of Immortality by Robert C. W. Ettinger; Copyright 1964 by R. C. W. Ettinger. Reprinted by permission of Doubleday & Company, Inc.

The Language of Life by George W. Beadle and Muriel Beadle; Copyright © 1966 by George W. Beadle and Muriel Beadle. Reprinted by permission of Doubleday & Company, Inc.

Acknowledgments

Can Man Be Modified? by Jean Rostand, translated from the French by Jonathan Griffin, *Peut-On Modifier L'Homme?;* ⓒ 1956 by Librairie Gallimard. English translation ⓒ 1959 by Martin Secker and Warburg Ltd., Basic Books, Inc., Publishers, New York.

Baby Want a Kiss by James Costigan; ⓒ 1965 by James Costigan. Reprinted by permission of Ashley Famous Agency, Inc.

Remembrance of Things Past, Swann's Way by Marcel Proust, translated by C. K. Scott Moncrieff; Copyright ⓒ 1934, 1962 by Random House, Inc.

"Artificial Intelligence" by Dr. Marvin L. Minsky, September 1966; Copyright ⓒ 1966 by Scientific American, Inc.

"The Mortal Cell" by Joan Woollcott, September 1967. By permission of Medical Affairs.

"White Collar Pill Party" by Bruce Jackson; Copyright ⓒ 1966, by The Atlantic Monthly Company, Boston, Mass. Reprinted with permission.

"Is Marriage Still Sacred?" by Ardis Whitman; Copyright ⓒ 1967 by McCall Corporation.

"Moral Problems in the Use of Borrowed Organs, Artificial and Transplanted" by J. Russell Elkinton, M.D. From editorial appearing in February 1964 issue of *Annals of Internal Medicine*.

"The Shape of Medicine in 1990" by Dr. Oscar Creech, appearing in *Medical World News* on Nov. 25, 1966.

Excerpts from four articles published in *The Journal of the American Medical Association*: "The Hopeless Case" by Dr. Frank J. Ayd, Jr., Sept. 29, 1962; "Life or Death by EEG" by Hannibal Hamlin, Oct. 12, 1964; "Rapid Scientific Advances Bring New Ethical Questions" by Dr. Eugene D. Robin, Aug. 24, 1964; and "Life or Death—Whose Decision?" by Dr. William P. Williamson, Sept. 5, 1966.

Excerpts from The Deerfield Foundation Lecture delivered by Joel Elkes, M.D., The Johns Hopkins University School of Medicine.

Excerpts from the James Arthur Lecture on the Evolution of the Human Brain called "Evolution of Physical Control of the Brain" delivered by Dr. Jose M. R. Delgado, 1965. By permission of the American Museum of Natural History.

Excerpts from a set of mimeographed lectures on ethics, Yale Medical School. By permission of Yale University.

Excerpts from *The Scientific Endeavor* by Dr. Neal Miller, 1965. By permission of the publisher, The Rockefeller University Press.

Excerpts from a Public Broadcasting Laboratory broadcast on medical morality, by permission of PBL.

Contents

And the Lord God formed man of
the dust of the ground, and
breathed into his nostrils
the breath of life; and man
became a living soul.

Genesis

When the dust of the earth
became conscious of the dust,
a transformation began to take
place in the face of the earth.

CHARLES AND MARY BEARD
The Rise of American Civilization

Foreword

Coming: the control of life. all of life, including human life. *With man himself at the controls.*

Also coming: a new Genesis—The Second Genesis. The creator, this time around—man. The creation—again, man. But a new man. In a new image. A whole series of new images. What will the new images be?

They will have to be quite different from the images we have known—the images that have led us to Vietnam, to turbulent racial conflict, to nuclear confrontation, to the threat of a polluted and overpopulated planet.

But all these things have come about—have they not?—with man at the controls, more or less. If he is acquiring awesome new powers—and he is—with immeasurably greater controls, does this not accelerate us all, at uncountable G's, toward the inevitable Dead End—a hundred bangs, followed by three billion whimpers?

If we believe so, yes.

But man's new images may offer us surprising alternatives. Their creation will require the energetic projection of the best minds of the race to the farthest reaches of their imaginations.

Futile effort?

Anything but. What we believe *about man, what we* want *for man, will profoundly influence what actually happens to man.*

What *do* we believe about man?

A fantasy of the future—a perhaps not so fantastic fantasy of the perhaps not impossibly distant future—may help crystallize the issue:

Imagine a dictator with a subject population—the dictator, a man who is sure he knows what is best for everyone; for himself, absolute power, for his subjects, happiness. He has at his command all the electrochemical techniques necessary for controlling the human brain as well as the most advanced methods

3

for controlling human reproduction. He can have entire populations raised "artificially" without resort to sex or family structure. He can also, if he chooses, have electrodes planted in the brains of his subjects, or begin administering "mind drugs" routinely, at a very early age.

This done, he can maintain his subjects in a state of hardworking subservience—constituting, in fact, a slave labor force—and at the same time keep them in a state of constant euphoria by stimulating the pleasure centers of their brains. Practically no one in such a society would have any true freedom of choice in any area of life where we now consider free choice important. But everybody would be happy.

Should anyone care about the *pursuit* of happiness if everyone already has it? If everybody is happy, can anything be wrong?

If we do think something is wrong with this happy picture—if we somehow find it profoundly depressing instead, is this merely due to the prejudice instilled by our own culture? Or is "happiness" perhaps *not* the goal of human life? Or, if it is, how do we define it? Can we answer the question persistently put to us by among others, Sir Julian Huxley: What are people for?

The answers (there are more than one) to such questions cannot be put off until the day of those future fantasies. We need to find some better answers now, even in the context of our more prosaic contemporary realities.

Not long ago a wealthy Southern shipowner, fatally injured in an accident, was rushed to the hospital. Though he was presumed dead on arrival, a team of doctors put forth heroic efforts to get his heart beating again. They kept it going, weakly and erratically, for some forty minutes. When further efforts failed, they finally pronounced him dead. Meanwhile, during the same critical forty minutes, a baby girl was born to the shipowner's only daughter. The daughter had married against her father's wishes. As a result, he had disowned her completely, leaving his entire estate to his son. But he did set aside $100,000 for any child of the daughter's who might be born before his death.

Was the new baby entitled to the inheritance? Was her grandfather dead when he got to the hospital, or was he alive?

Certainly, by all traditional standards, he was dead. And yet—doctors these days do frequently succeed in resuscitating patients who, in the very recent past, would have been considered quite irrevocably dead. They labor mightily, restore the heartbeat, and by and by the corpse is up and smiling. If the shipowner had got up smiling, he would have been indisputably alive. This being so, could he really have been dead on arrival? If so, would this have meant that he had been brought *back* from death?

This particular case was settled out of court, so no judge or jury had to grapple with these questions. But consider another case that did get to court. A wife, separated from her husband, had already been awarded custody of their child. Now she was trying to deny her husband visiting privileges. He had no rights, she said, because the child had been conceived by artificial insemination, the semen having been obtained by the family physician from an anonymous donor. The husband protested vehemently. The insemination had taken place with his full knowledge and consent, and the child had been born in wedlock. He had, like any husband, shared the joys and worries of his wife's pregnancy and awaited word from the delivery room with the same anxiety as any other expectant father. He had loved and nourished the child as his own, and he considered himself the child's own father.

The court granted him visiting privileges. But when the

mother moved from New York, where the decision had been handed down, to Oklahoma,* where she reopened the case, the court there saw the matter in quite a different light, ruling that her husband had no parental privileges, since he was not in any sense the child's real father.

If he was not the child's real father, who was? The anonymous stranger who donated the semen? If so, could he now claim the child as his? Could he demand visiting privileges? Conversely, could he be forced to support the child? Is the child his heir? Could the mother's husband, when the child was born, have had the right to disown the child and declare him illegitimate?

None of this is frivolous conjecture. These are real questions arising from problems that trouble real people. Similar perplexities arise every time biology or medicine makes an important advance. The advance does not have to be a major breakthrough in basic knowledge. It can be a relatively simple technique, such as learning to restart a stopped heart, or to inseminate a woman artificially. Once the discovery is made, once the technique is learned, once it can be applied with reasonable safety, those who feel the need for it tend to apply it. But there is always a painful lag before the mechanisms and attitudes of society, including the law, catch up with the new reality that science has wrought. As man's power to control life accelerates, this kind of lag will be more than merely painful; it could be catastrophic.

The lag has been tolerable until now only because the problems so far raised are almost childishly simple and straightforward compared to the braincracking complexities—legal, social, ethical, moral, philosophical, religious, esthetic, political—which are soon to be thrust upon us. As science goes its

---

* Oklahoma recently became the first state in the U. S. to legalize artificial insemination.

headlong way, the lives of men will undergo transformations so sweeping as to constitute, for all of us, a wholly new world, a world without precedent in human history.

"One thing that is new," said the late Dr. J. Robert Oppenheimer in a lecture he gave on the occasion of Columbia University's bicentennial, "is the prevalence of newness, the changing scale and scope of change itself, so that the world alters as we walk in it, so that the years of man's life measure not some small growth or rearrangement or moderation of what we learned in childhood, but a great upheaval." Oppenheimer's lecture was delivered in 1954—in the olden days, scientifically speaking, way back before the first sputnik was put into orbit, back before the biochemists made their monumental breakthroughs in cracking the genetic code. The world is changing at an even more breakneck pace, and the attendant dilemmas will soon be crowding in upon us.

To describe the radically metamorphosed world of the near future, scientists have had to grope for new words. Sir Julian Huxley suggests that the new man be called "trans-human." Dr. Kenneth Boulding believes the new civilization will represent such a leap ahead that it ought to be called something like "post-civilization." Dr. Harlow Shapley calls it the Psychozoic Kingdom. The late French Jesuit scientist-philosopher Père Teilhard de Chardin, coined a whole new terminology for the future.

"Whatever terms we use," says Dr. Jonas Salk, physician, scientist, and biophilosopher, "the time has come when the public must be made aware of the great impact of biological thought and knowledge. Such awareness will lead to the further liberation of man and the flourishing of his great potential. But he will be called upon to avoid the new dangers that liberation brings. Here only his sense of values can be his guide. Meanwhile the managers of society should be given some advance notice of what is in store."

7

Foreword

The account that follows will not be in any sense an all-encompassing survey of the startling and audacious investigations now being pushed by biologists and medical scientists, nor will it provide completely detailed descriptions of any one of these. Rather, this is an attempt to give "some advance notice of what may be in store" for mankind as a consequence of current research, discovery, and achievement in a number of different scientific fields—especially those related to man's new abilities to tamper with* his body, and therefore with his psyche. Not that there is anything new about man's attempts, many of them successful, to tamper with his body. What else is the history of medicine all about? We tamper, in a sense, when we take an aspirin or an antibiotic, when we submit to an appendectomy or a tonsillectomy, even when we go to the dentist or do calisthenics.

But the latest ways of tampering are radical departures from the ways we have known. They involve such procedures as the wholesale replacement of failing body parts with transplanted or artificial organs; the control of the body, brain, and behavior through electronics, drugs, and cybernetics; the freezing of dead bodies for possible earthly resurrection by the increasingly sophisticated science of the future. Some procedures involve tampering before birth—in the prenatal period, for example, while the fetus is still in the womb—and a variety of methods

---

* I use the phrase "tamper with" to mean any interference by man with the workings of the body; anything that alters the course of what might have happened had he not intervened. The outcome can be beneficial or deleterious. The word "tamper" often carries negative overtones, connoting a change for the worse, and I suppose the reason I use it here is that medical intervention, though intended for at least the long-run benefit of patients, often does carry overtones that warrant some alarm. The scientific tamperer or tinkerer is usually fascinated with things and ideas in themselves and how they work; the consequences for people—good or bad—are not always uppermost in his mind.

for conceiving and growing babies inside or outside the womb, with or without intercourse, not to mention the further possibility of modifying future generations through eugenics or through the actual molecular manipulation of the genes.

Scientists are daily edging closer to the deciphering of nature's subtlest biochemical codes, and each new discovery provides yet another startling means for fiddling with our bodies and our brains. As a result, no idea seems too wild to contemplate. What would you like: Education by injection? A catalog of spare body parts? A larger, more efficient brain? A cure for old age? Immortality through freezing? Parentless children? Custom-ordered body size and skin color? The ability to convert sunlight directly into energy, just as plants do, without utilizing food as an intermediary? Name it, and somebody is seriously proposing it.

In 1965, Dr. Charles C. Price of the University of Pennsylvania, then president of the American Chemical Society, said flatly, "The synthesis of life is now within reach." He proposed that the United States make the creation of life in the laboratory a national goal, comparable to sending astronauts to the moon. In sober scientific circles today, there is hardly a subject more commonly discussed than man's control of his own heredity and evolution. And the discussions seldom leave much doubt that man will acquire this control. It is a matter of when, not if.

Scientists tend to agree that some of the most exciting future developments will come out of insights and discoveries yet to be made, with implications we cannot now foresee or imagine. So we live in an era where not only does anything that we can imagine seem possible, but the possibilities range *beyond* what we can imagine. In such an era it is hard to tell physics from metaphysics, to distinguish the fictional "mad scientists" from the sane nonfictional ones, to judge what is a true possibility and what is sheer rot. But there is no resolving this kind of uncertainty. Scientists may deplore the more extravagant extrapolations of the journalists—but it is hard to exceed the scientists' own more uninhibited speculations. And they are the first to admit that even they cannot give us sure guidance as to what is really going to happen.

There are powerful institutions to give us guidance about what *ought* to happen—the most powerful, perhaps, being religion. Regardless of what science makes possible, moral approval or disapproval has, throughout man's history, influenced which advances he accepts instantly, which he accepts more slowly, and which he rejects altogether. In the new age, however, it is unlikely that any advance can be totally ignored.

Physicist Johannes M. Burgers, when he retired from the University of Maryland, was so alarmed about experiments that might lead to man's duplication of himself in the laboratory that he proposed a fifty-year moratorium on work toward this end. He said, "It may sound anti-progressive. I would not want to make it forever. But for the moment, wait. Wait until the educational level of man is higher. Wait until you know more about life." Lewis Mumford has similarly proposed "a deliberate act of restrictive self-discipline on the part of science and technics." But scientific curiosity is one of the strongest motivating forces in the world today, and the prizes for the pursuit of that curiosity are often spectacular.

Some scientists in some countries will follow any line of research (probably pursued for good ends, to begin with) that happens to fascinate them, regardless of prevailing moral attitudes. Some of the discoveries and achievements will, at first, have only a limited impact on only small and restricted segments of the world's population. But the powers and advantages that can accrue to those who exploit them will be so overwhelming that the rest of the world will not long shut them out. And those who guide us, including the theologians, will not be able to guide us truly without taking them into account.

The more dire implications of biomedical discoveries are considered by some to be too futuristic to concern ourselves about more than casually, at a time when there are such overridingly immediate problems to occupy us: survival in the nuclear age, the population explosion, race relations. But it would be a mistake to view the biomedical predicaments as lacking a sense of urgency. "These are *not* long-term problems," says Dr. Joshua Lederberg, Nobel biologist at Stanford University. "They are upon us now." As man's knowledge—and therefore man's power—takes on new dimensions, hardly any human concept or value will remain too sacrosanct to undergo a wrenching reappraisal. Health and disease, youth and age, male and female, good and evil—all these will take on transformed meanings. Life and death will have to be redefined. Family relationships will perforce be quite different, and even individual identity may be hard to ascertain. Nothing can be taken for granted among the trans-humans in the post-civilization of the Psychozoic Kingdom. To preserve any semblance of current human values, the leaders of men will have to think in new categories, and with creative leaps of the imagination.

In the fall of 1965, *Life* magazine published a four-part series called "The Control of Life" to help call public attention to the vast implications of biomedical advance—and to the sense of urgency advocated in scientific circles by people like Lederberg and Salk. In the intervening years these implications have been worried about aloud in an ever-increasing quantity of lectures, seminars, symposia, books, articles, sermons, radio and television programs. "Everywhere I go these days," says Dr. Richard Farson, director of the Western Behavioral Sciences Institute, "there is a session, either official or unofficial, on this topic. Where does man go from here, everyone wants to know, and what will human life be like in the year 2000?"

With contemplation of the biological future becoming such a universal pastime, it is time somebody gave the activity a name. I hereby offer my candidate: *biosocioprolepsis*. Prolepsis, according to *Webster's Unabridged,* means "anticipation; specif.: . . . *Rhet.* A figure by which objections are anticipated in order to weaken their force," prolep*tics* being "the art or science of prognosis." By projecting our imaginations ahead into our possible choice of social futures, we try to anticipate the dangers inherent in biomedical advance, and to forestall them by our foresight. Prolepsis will here be stretched to mean, as well, anticipating the hopes and benefits, in order to foresee ways to take advantage of them. *Bio* is for biology, *socio* for sociology. Biosocioprolepsis, then, is the anticipation of biology's impact on society. Let's call it BSP for short.

The hopes and hazards we anticipate in BSP are all based on the fact that man, who has already learned to remake his physical environment, will now acquire—or have thrust upon him—the capacity to remake himself. The dust of the earth, having become conscious of the dust of the earth, will be able to recreate itself without benefit of the original creator's breath —and to recreate itself in virtually any image, thus becoming an active participant in the new Genesis.

There is a volume of science fiction by Frederik Pohl called *The Case Against Tomorrow*—a title that could almost have served for this book as well. Almost, but not quite. I do make a case against tomorrow in that I conjure up a variety of uninviting prospects that tomorrow might encompass. The fact that I do conjure them up should not be taken to mean that I advocate them, any more than an epidemiologist's warning about the threat of a typhus outbreak means he is in favor of having people come down with the disease. Besides, the prospects are not all bad, and I hope that I shall also succeed in making a case of sorts in favor of tomorrow.

The kind of case we make, for or against, is of course academic. Tomorrow will arrive, with or without our approval. What kind of tomorrow? Well, we may have a lot to say about that. We don't have to let it just happen to us. If we are willing to undergo a few moderately strenuous exercises in BSP, we might then be able to equip ourselves with the knowledge and understanding to shape that inevitable tomorrow into the kind of time that we will be happy to dismiss the case against.

"Admittedly," says Dr. René Dubos in *Man Adapting,* "there is an element of intellectual conceit, and also of naïveté, in even attempting to postulate what would be a proper scientific course of action for the future welfare of mankind, because the future is so uncertain. But a defeatist attitude on this score would be justified only if human life were completely determined by blind forces. In practice, the future is the creation of man as much as it is the result of circumstances. True enough, the 'logical' future is the expression of natural forces and antecedent events. But there is a 'willed' aspect of the future, which comes into being to the extent that men are willing to imagine it and to build it."

"Man," says Dr. Shapley, "as half beast, half angel, must of course comply with the biogenic law, but he is able to make amendments thereto. It is probable, and certainly deeply to be desired, that the men of the future will correct our shortcomings and build on the basis of our thoughts and acts a finer mental and social structure."

The fruits of scientific adventure do not necessarily add up to unmitigated woe for man. Indeed, such new knowledge and power may help us overleap some of our old, seemingly insoluble woes. "Can the Ethiopian change his skin," asked

Jeremiah, "or the leopard his spots? Then may ye also do good that are accustomed to do evil." Thus, if it suddenly turns out that the Ethiopian and the leopard and you and I can change anything we care to change, then it follows—does it not?—that we may also do good. *Gaudeamus igitur.*

It would be fraudulent for me to pretend that the subject matter and ideas to be set forth in this book will always be uncomplicated or that the reading of it will be totally effortless. However, I have tried to keep the presentation as simple and as straightforward as possible, and the style strives to be lucid rather than lyrical. Nor is the content intended as a compendium of data in a field where the fantasies of fiction are rapidly becoming the biological facts of our time.

Part I of the book, "The Refabrication of the Individual," will deal largely with such subjects as resuscitation from clinical death and the prolongation of life through gadgetry, geriatric research, transplantable and implantable vital organs; in brief, those tamperings with the human body that are already being practiced or contemplated by the medical profession. These are therapies that might directly and personally affect almost anyone now living, and they raise dilemmas requiring immediate confrontation not only by doctors but by their patients and families. Part I will also touch on prospects such as body-freezing—eventualities which, though further off in time, if not altogether unattainable, are worth our current attention. The quandaries they give rise to, and the value judgments we must make, can shed valuable light on, and insight into, those projects whose immediacy force us to place them at stage center.

Part II, "The Exploration of Prenativity," will deal with biomedical tampering before birth—from the fetus in the womb all the way back to the sperm and the egg and their chemicogenetic contents. All of us now living are safely beyond this kind of tampering. But the control of prenatal life may quite possibly affect the children—and almost certainly the grandchildren—of many now alive; and the truly sweeping connotations of asexual reproduction and genetic manipulation are such that we would do ourselves a disservice to set them aside. We must have an unflinching look at them if whatever human values we arrive at for today are to serve us for very long.

The subject matter of Part III, "The Control of the Brain and Behavior," might well have been included in Part I, since the brain is not truly separable from the body. But I have arbitrarily separated it because it is, in my view, the most important and complicated aspect of the control of life as well as the one most fraught with hopes and terrors for man.

I apologize in advance for inaccuracies. I have, of course, tried hard to avoid them, but it seems likely that a few will have crept into a work so unavoidably arrogant in its scope.

PART I

The Refabrication of the Individual

Biomedical research has traditionally pursued one primary goal: helping the sick—with emphasis on finding remedies for their ailments, or at least relief from their suffering. Medical scientists have not lost sight of this major impetus, nor do they plan to. But an increasing proportion of their energies has lately been devoted to a quest for understanding life's basic processes, and especially its control mechanisms.

Scientists closely study individual cells—from the tiniest microorganisms to human body cells in all their variety—in an attempt to puzzle out their complex workings. They have been focusing special attention on the individual molecules within the cells—those that constitute the cell's own substance as well as its transient passengers—and they suspect that even the quantum-mechanical activities of the atoms and electrons that make up those molecules, as well as their precise placement in the molecular configuration, have great biological import.* They study the body's organs and systems, separately and in synchrony. They study the whole body as an integrated, dynamic, delicately balanced entity, in disease and in health, not overlooking the effects of the psyche, the "mind," the emotions, and the cultural outlook on the physiological functions, and vice versa. They study the impacts, subtle and gross, of the physical environment on the human organism, and note the biomedical effects of the interactions of people with one another.

The research spectrum stretches from purely theoretical considerations to *in vitro*** laboratory science to animal experimentation to clinical tests on human beings to statistical and

---

* This is not to say that it will necessarily ever be possible to reduce or explain biological systems strictly in terms of physics or quantum mechanics.

** *In vitro* means "in glass"—referring to test tubes or other laboratory glassware, as distinguished from *in vivo*—in a living organism, usually an experimental animal.

epidemiological studies of entire populations. As their labors begin to bear fruit—that fruit off the tree of knowledge, no longer forbidden but eagerly sought—and as the new knowledge is collected, synthesized, digested, and acted upon, men will be in a position not merely to combat their illnesses more efficaciously, but to attain a more positive state of health and well-being, along with an enhanced capacity for awareness and joy, and prospects for a longer, and hopefully a happier, life.

With a lot of luck, that is.

Every new piece of knowledge is a tool as well as a weapon. It gives hope, and it carries risks. A knife cuts bread—or stabs enemies (or friends, for that matter). Nuclear fission and fusion can provide energy for man's work, or nuclear and thermonuclear bombs. In the same manner, no therapy is without its concomitant dangers. To apply the full benefits of medical knowledge requires true understanding, which in turn implies wisdom—never a surplus commodity. The odds are always high that men will steadfastly maintain their perverse inability to exercise their intelligence in their own self-interest. Besides, as Dr. Dubos has suggested so convincingly in *Mirage of Health,* disease and danger may be the price of life itself, and the attainment of perfect health an illusion. Even so, we must strive for perfection if we are to come anywhere near it. A man's reach, as Browning said, must be farther than his grasp.

If we learn with certainty that substances in the environment are deleterious to health, we can hope to find ways to remove or modify them—and even convince people that what needs to be done ought in fact to be done. If we learn that certain attitudes can be harmful to the individual, we may be able to change his attitudes—or at least let him know the consequences and offer him a choice of options (in which case, of course, he may in all good sense and conscience prefer the consequences to the cure). If we learn the causes of an ailment, we may reasonably hope for a cure. If we understand a biological

20

process in detail, we can perhaps find ways to interfere with it, speed it up, slow it down, or reverse it. And if we learn, with precision, the control mechanisms of life—why, then, the controls pass to us, if we choose to exercise them—a choice we must consider seriously before we back away from it, lest someone exercise them on *us*.

One barrier to the successful practice of BSP* is our hardwon empirical understanding that the consequences of scientific progress often tend to be unforeseeable. It is a cliché to point out that a basic advance in fundamental scientific knowledge nearly always pays off in terms of practical technical applications. What is less well understood—or at least less frequently articulated—is that the reverse is also true: a new laboratory tool or technique can pay off in terms of basic scientific knowledge. It does so by opening the way to new exploration and experimentation. Any number of such tools or techniques might be cited in the biomedical field—the electron microscope, for instance, or the use of radioactive tracers. But the technique I shall dwell upon at some length as a prize example of what I am talking about is tissue culture. I have chosen tissue culture because some understanding of its power is essential to the understanding of many of the other topics I plan to discuss in the course of this book.

When Dr. Ross Harrison of Johns Hopkins University first succeeded in keeping alive a bit of frog nerve tissue in a culture of the frog's natural body fluids back in 1907, he was doing it to settle a point of controversy as to whether or not nerve fibers grew from single cells. (They did.) But what excited the medical world—including researchers with no special interest

---

* In the belief that, whenever an author coins a new term, he ought to define it more than once: BSP stands for biosocioprolepsis, or the anticipation of biology's impact on society—and the application of foresight with a view to forestalling the evil and taking advantage of the good.

in the nervous system—more than Dr. Harrison's solution of the controversy was his incidental demonstration that tissues could be kept alive outside the body. Soon a whole variety of tissues were being kept alive and growing (i.e., the cells absorbed nutrients from the surrounding medium and were able to reproduce themselves) in a whole variety of culture media: tissues from several species of birds and mammals, including man. Thus, many biomedical scientists in many nations became cell farmers, either as a specialty or as an adjunct to their main line of research. The most celebrated of these husbandmen of cells in the decades preceding World War II was the Rockefeller Institute's Dr. Alexis Carrel, who was celebrated for a number of other good reasons as well—among them the introduction of new methods for suturing blood vessels, his pioneering experiments in organ transplantation, and the development, with Charles A. Lindbergh, of a "mechanical heart." One of Carrel's more impressive demonstrations of the potential of tissue culture was the long-run performance of a sliver of chicken heart he put in a flask in 1912. The chicken-heart tissue was kept alive and growing for the remaining thirty-three years of Carrel's life—and outlived its author by another two years. Despite his brilliant achievements as surgeon and experimenter, despite his Nobel Prize, Carrel became so eccentric, and surrounded his research with such an aura of theatricality (he made his lab technicians wear flowing black robes and hoods) that he gave tissue culture a bad name, and it went into an unfortunate period of decline.

What gave tissue culture new life was a transfusion of money in the late 1940's. The transfusion came about because the National Research Council and the American Cancer Society wanted to test chemicals on malignant cells, and they saw tissue culture as the easiest way to accomplish it. The successful chemotherapy of cancer requires that the given drug attack or kill cancer cells while leaving normal cells relatively unharmed. In tissue culture, drugs can be tested endlessly without risking anyone's life, and the results can be observed and measured on both malignant and normal cells in a controlled laboratory environment. So NRC and ACS jointly sponsored a national conference of the country's leading tissue-culture experts in June 1946, and soon began issuing research grants. At the time of the meeting barely a dozen specialists could be brought together. Seven years later the newly-formed Tissue Culture Association could count 300 qualified members. Since then the membership has more than quadrupled.

Tissue culture can make a strain of cells virtually immortal. Take the case of Mrs. Helen L., a lady in her thirties who died of cancer of the cervix in the Johns Hopkins Hospital in 1951. After her death, scientists were able to keep her cancer alive, for study, in tissue culture. Dr. George O. Gey named the culture the HeLa strain, using the first two letters of the patient's first and last names. The malignant cells multiplied with astonishing rapidity compared to normal cells, and they proved to be unusually stable—that is, they retained their characteristics unchanged through many generations. Because of this the HeLa strain has for many years been used as a standard in tissue-culture studies, and test tubes of it have been shipped out by the thousands to laboratories all over the world. Because of a further discovery, made by a University of Minnesota team, that the HeLa strain exercises an extraordinary attraction for polio viruses, it has been used as a standard tool for detecting the presence of polio viruses.

Tissue culture played a prominent role, not only in the detection of the polio virus, but in the conquest of the disease itself. Crucial to Dr. Salk's success in speedily producing a usable vaccine was the discovery, made by Dr. John Enders and his associates at Harvard, of how to grow polio viruses in a proper tissue-culture preparation. At a meeting in Copenhagen in 1951, where Enders reported his new method, Salk reported on the advantages of it. "We have been able," he said, "to prepare from the testes of a single monkey 200 tubes which, for immunologic studies, would correspond to 200 monkeys, using the same number of tubes as one would use animals."

Employing tissue-culture techniques, scientists can singe cells to try out burn treatments, freeze them to learn about frostbite, irradiate them to see what damage is done and how it might be repaired, saturate them with smog or pollens to study allergic effects on respiratory-passage cells. Out of tissue-culture studies have come not only new drugs and vaccines, but altogether new therapeutic concepts. When the Dutch scientist Dr. Peter Gaillard saw, for instance, that parathyroid cells grown *in vitro* continued to manufacture their characteristic hormones, he followed up his discovery by implanting colonies of the cells into victims of parathyroid deficiency. In patients under thirty, the implants took hold, and the deficiencies (usually fatal) were cured. Gaillard's work stirred much experimentation, and much speculation about the future of hormonal implants, all the way from birth-control compounds to permanently implanted insulin-producing cells for diabetics.

Many of the scientists specializing in cell agriculture do so, not with any specific therapeutic goal in view at all, but merely to take advantage of the opportunities tissue culture provides for the study of the cell itself.

One of the most fascinating parts of the cell is its nucleus, the seat of control and power where the chemicals that govern life reside. The basic genetic material, the life-governing chemical substance, is deoxyribonucleic acid or DNA—a once-esoteric set of initials now just about as familiar to the American household as PTA or DDT. Though DNA had been isolated as early as 1869, it was not until the 1940's that a series of experiments at the Rockefeller Institute proved it to be indisputably the principal chemical of heredity, and it was only in the 1950's that Dr. Francis H. C. Crick and Dr. James D. Watson at Cambridge elucidated its architecture. Their epic illumination, and the explosion of research it inspired, have shown that in the coiled, spiral-staircase, double-helical structure of the DNA molecule, and in the complex arrangement of its atoms, lie the key secrets of heredity, of human development, of aging—and perhaps of mind and memory. DNA's genetic messages are written out in a four-letter code, each "letter" being a specific chemical compound or base unit of the DNA chain. Using its code of four chemical "letters" (adenine, thymine, guanine and cytosine), DNA transmits its instructions via RNA, another nucleic acid similar to itself in structure, though RNA (which comes in several varieties) is a single strand rather than a double helix, and cannot duplicate itself. RNA also uses a four-letter chemical code, but one of the chemicals is different; in its alphabet, uracil replaces thymine.

This was not too much information to go on, but it was enough to open up the possibility of cracking the genetic code. The first great stride in that direction was made in 1961 by a young biochemist at the National Institutes of Health, Dr. Marshall Nirenberg. His historic experiment is worth an appreciative look, since it provided the biological Rosetta Stone, the crucial key to translating the hieroglyphics of genetics.

Nirenberg and his associates found the key they were seeking in the manufacture of proteins, one of the cell's most important functions, a task carried out largely by RNA under orders from DNA. Proteins are made by putting together their building blocks—amino acids. If RNA makes proteins, it must do so, Nirenberg reasoned, through very explicitly coded instructions. Somehow its four letters, in one combination or another, must recognize which specific amino acids to select to build which specific proteins. The proper combination of letters would be the code word for a given amino acid. Such an amino-acid code word would have to contain more than two

25

letters because there would not be enough two-letter words to go around for all twenty amino acids. Three-letter words, however, ought to work just fine. But which three-letter words, out of sixty-four possible combinations, would select which amino acids?

It was as if there were a pack of twenty dogs whose names you had to learn. Though you don't know their names, you do know that all the names are different, and that each dog will answer to his own name and no other. All you can do, then, is start thinking of possible dogs' names and calling them out. It you holler "Rover" and nothing happens, you try again. If you holler "Fido" and one comes running, then there are only nineteen names left to guess.

This is the same sort of technique Nirenberg used. First, he mixed up a batch of chemicals prepared from living cells, including all the amino acids. He made sure that his precise recipe contained no RNA of its own to foul up the experiment. Then he manufactured an artificial RNA, made up entirely of only one of its four code chemicals—uracil, which can be thought of as the letter U. The only three-letter word that can possibly be made out of U's is UUU, and the sequence doesn't matter since, however you scramble it, it still comes out UUU.

Adding this artificial RNA, made of nothing but U's, to the amino-acid soup, Nirenberg got a protein made entirely of a single amino acid, phenylalanine. There could be no mistake—he had talked to the amino acid in its own language. Instead of "Fido," he had hollered "UUU," and phenylalanine had come running. It was clear that UUU meant phenylalanine. In the new molecular dictionary, phenylalanine was the first word to be translated from the genetic.

Other scientists quickly joined Nirenberg, hollering "Rover" and "Sport" and all the other names they could think of. Their hollering, of course, consisted of delicate and painstaking chemical operations, often using radioactive tracers to signal the whereabouts of specific substances. In their work they had to deal with complications which did not concern Nirenberg in his pioneering undertaking. For one thing, they had to make their artificial RNA out of more letters than merely U. For another, in words with more than one letter, the sequence begins to matter—cta does *not* spell cat.

Despite these difficulties, scientists have gradually come to know the code words—or, at least, the letters in the code words—for all twenty amino acids. They have also learned that sometimes there is more than one word for the same amino acid—just as "car" and "auto" can convey the same message in English. This gives the organism a better chance to survive. If one of the key code chemicals were knocked out, or in short supply for any reason, the amino acid would have an alternate choice, and protein manufacture would not necessarily come to a stop.

Another discovery, among the many that followed, was the existence of genetic punctuation. The DNA molecule is not a single, uninterrupted chain full of nothing but the code chemicals. Every so often the genetic messages are interrupted by punctuation marks. Instructions for how to make a specific protein may take up, say, a thousand rungs in the DNA ladder. At the end of this set of instructions there is a bit of protein which acts as the period at the end of the sentence.

In record time since the Nirenberg experiment, scientists have made enormous strides toward being able to read the genetic code. And the code turns out to be universal for all the living world.* The combination of base units that dictates the

---

* This conclusion is still controversial as of this writing. For instance, the same set of data from the same set of experiments that led Dr. Nirenberg to this conclusion, lead Dr. Barry Commoner to the opposite conclusion. Most convincing to me were a series of experiments done by Dr. Philip Handler and his associates before he left Duke University to become president of the National Academy of Sciences. As reported

manufacture of a given protein, dictates the same protein in germs, snakes or men—and did as much for the dinosaurs.

Scientists have already succeeded in building up DNA from its constituent chemical building blocks in the laboratory. They have induced genetic mutations in fungi and fruit flies by radiation, and, by mixing the DNA from one species of bacteria with that of another, have produced permanent hereditary changes. They have, in addition, been able to pin down some genetic defects in human beings—such as sickle-cell anemia and the form of mental retardation known as phenylketonurea (PKU)—to the presence or absence of a specific chemical substance, or even to a single small segment of that substance. Most of this has been accomplished through relatively rough-and-ready methods which are in process of being rapidly refined. Though the science of molecular biology is only in its infancy, it is an infant like Gargantua, with the strength for mighty feats; it is an infant like Davey Crockett, killing b'ars with its bare hands.

What really distinguishes molecular biology from all biology before it is that it *does* penetrate down to individual molecules. Biologists were once limited to the study of gross organisms, or pieces of them. They gradually learned to deal with individual cells—usually dead ones, artificially stained, observed under the microscope. Tissue culture has made the cell accessible in its living state, alterable almost at the convenience of the experimenter. But even an individual cell is a fairly gross and complicated object. At any given instant, it is dependent on the simultaneous carrying-out of myriads of chemical reactions

---

in the *Journal of Biological Chemistry,* Handler and his team selected a key genetic enzyme, phosphoglucomutase, and were able to isolate it from man as well as several species of animals and plants: rabbit, shark, flounder, oysters, chicken, brewer's yeast, potatoes, sweet potatoes, and three varieties of bacteria. In all cases the enzyme performed the same biochemical conversion—at least, in the test tube.

involving not only thousands of molecules, but thousands of different *kinds* of molecules. So getting down to the manipulation of those individual molecules represents a truly revolutionary advance.

To appreciate its nature, imagine a Martian astronomer who has been studying life on earth. He has done the best he can do from way out there, observing us through the best instruments at his command—just as the biologist has done the best he can do, looking through his microscope. All that the Martian could observe, and then only in the vaguest way, were the mass activities of entire populations. He had no idea that a Napoleon or a Newton might exist or how his deeds might affect the destinies of millions. What he knows about human behavior is understandably limited.

But suddenly, through some remarkable new techniques and instruments, he finds out about individual people. He learns that they communicate and order their lives with the help of a language. More than that, he figures out all the letters of their alphabet. Now, if he can only learn to spell with the letters . . . you can imagine his excitement. You can see how rapidly his knowledge of life on earth would begin to proliferate.

Biologists on earth have arrived at exactly this stage. Heretofore they could observe only cells. A cell is not an individual, as Dr. Salk has often pointed out, but a community—a highly organized community in which a vast and diverse population of molecules go about their business. Now that the biologists have begun to observe and manipulate the molecules—the real individuals in the cellular population—they are even more excited than our imaginary Martian astronomer, for what they are studying is not life on another planet, but the life pulsing in their own bodies. The quest is bound to be pursued vigorously. If they really learn to read, in explicit letter-by-letter detail, the genetic data written out in the DNA molecule, then it follows that they might learn to modify or edit the instructions, or to

29

write out genetic messages of their own. The potential control of the genetic material is certainly of paramount importance to every medical question still open. In fact, it reopens many medical questions considered closed.

Several years ago, I was driving from New York to Philadelphia with my family in our old Chrysler, which had served us beautifully, with no major troubles at all, for many years. But that Saturday morning, on the New Jersey Turnpike, the engine went out, all at once. It simply quit, like the wonderful one-hoss shay, and was good thereafter only for the junkyard. At the time, I couldn't help thinking: That's the way I'd like to go. With all cylinders going to the end. To feel, like Walter Savage Landor, that "I warmed both hands before the fire of life; It sinks, and I am ready to depart."

The great nineteenth-century biologist Dr. Elie Metchnikoff had the same kind of yearning, and he wondered why so few people seemed ready to go when the time came. He became obsessed with the idea, and decided it was because men usually die before they attain the life span of which they are truly capable. His desire to increase human longevity led him, in his later years, into a series of inanities. (It is easy to call them inanities long after the fact. If the inanities had turned out to be marvelous ideas, as inanities sometimes do, we would now be calling them marvelous ideas.)

Every man learns early that his days are numbered. But he hopes that the number will be high. "Aging," as British biologist Dr. Alex Comfort puts it, "is an unpopular process with man." It is not merely unpopular with individual men, but is a major tragedy for society. Just as a man finally reaches a point in his chronology where his painfully accumulated store of wisdom and experience will perhaps enable him to make his major contribution to his fellow men; just at this same point where he has finished raising his children and discharged his basic social obligations, leaving him free to concentrate on whatever creative longings he has postponed; precisely at this point, his energy declines and his organism begins to deteriorate. Not only does a man lose these years, but we all lose the potential products of them. No wonder George Bernard Shaw complained that youth was wasted on the young.

In his wish for longevity man relies heavily on medical science; and his reliance has not been misplaced. Medicine has always spent its best efforts on the postponement of death, and has succeeded so brilliantly, with its triumphant parade of disease-defeating drugs, therapies, gadgetry, vaccines, and surgical feats, that the average life expectancy at birth in the United States has gone up from forty-seven in 1900 to the biblical three-score-and-ten in 1968.

But this leap in life expectancy is misleading. It is a statistical fact which means only that the *average* life span has increased. That is, by virtue of a substantial decrease in infant mortality and the conquest of a number of infectious diseases that used to kill off large quantities of people at an early age, *more* of us are living into old age. In Civil War days, a scant 3 percent of the American population was over sixty-five; today the proportion is more like 10 percent. In the last century the population over sixty-five has tripled, and the middle-aged have doubled—not in mere numbers, but in proportion to the rest of the population. But in terms of senile degeneration, an old man today is still as old as an old man used to be. There are medical palliatives available to make life more comfortable for the old folks—a whole medicine-chest full of drugs, ranging from tranquilizers to anticoagulants to sex hormones, as well as surgical techniques and anesthetics which make possible operations on elderly patients who once would have been classified as poor risks. But no one has yet held up his divining rod and found the rejuvenating waters of the Fountain of Youth. Not that this has been medicine's serious goal, anyway. What scientists do

31

seriously hope to bring about is expressed in the motto on the masthead of the *Journal of Gerontology:* "To add life to years, not just years to life."

To live long, but not to age; to endure and still retain one's vigor and intellectual faculties; to be able to say, with Rabbi Ben Ezra, "Grow old along with me!/ The best is yet to be./ The last of life, for which the first was made . . ." Americans at least as far back as Benjamin Franklin dreamed this same dream, and gerontologists* have adopted it as their own—but so far it remains just that: a dream. If old people in earlier days seemed heartier, it is because they were. Those who survived the rigors of life *had* to be innately tough, or unusually lucky, to make it all the way. The vigorous and productive octogenarians and nonagenarians of our own day—the Churchills, the Adenauers, the Schweitzers, the Picassos, the George Bernard Shaws and the Bertrand Russells—retained their powers not through any medical intervention, but by virtue of the sound constitutions they inherited. For the most part, we are keeping people alive longer by providing a great variety of medical props. Many are becoming what Dr. Edward I. Bortz, a member of the AMA's committee on aging, has called "chemical Methusalehs"—with implications that already plague society. Dr. Comfort, himself one of the world's outstanding gerontologists, likens these vegetating senior citizens to Tithonus of Greek mythology—Tithonus, to whom Aurora gave immortality but forgot to include a proviso for youthful vigor—Tithonus, doomed to eternal decrepitude, finally praying for death.

There is, of course, good reason to hope that we will fare

---

* The disciplines of *geriatrics* and *gerontology* overlap in many areas. But, to make a rough general distinction, *geriatrics* is concerned mainly with the medical care of the aged, with the diseases of aging, while *gerontology* mainly studies the aging process itself, in the long-range hope of finding ways to modify it.

better in the future. "Gerontology aims," says Comfort, in *The Nature of Human Nature,* "to increase the longevity of the species by actually slowing down aging. Whether it will be possible to do this, how large a change we may hope to produce, and by what means we might produce it are still not known, but with 600 teams clocking on daily in the U. S. A. alone to work on problems connected with aging, and growing activity in other countries, results should soon appear."

How soon? Dr. H. Bentley Glass, the distinguished Johns Hopkins biologist who recently became provost of the State University of New York at Stony Brook, predicted in a recent speech that, by the year 2000 most people will remain vigorous until the age of ninety or one hundred—just as Metchnikoff hoped they might.

In the opinion of Dr. James Bonner of Cal Tech, "biologists are on the verge of finding a way to eliminate senility, thus facilitating a human life span of 200 years." Many scientists would decry these estimates as too optimistic, and express skepticism about what "on the verge" means in Bonner's forecast.* Regardless of timetables, recent studies in aging, including Bonner's own, make it evident that grounds for new optimism do exist. It is no longer fatuous—or even "journalistic" (that favorite cussword of academicians)—to suggest the attainment of such life spans at some point in the not-impossibly-far-off future.

---

* Dr. Jean-Bourgeois Pichat, head of France's National Institute for Demographic Studies, predicted in 1966 that within fifty years some people might be able to live essentially forever—basing the prediction on the assumed acquisition of knowledge that would allow treatment to begin at birth. In June 1967, Dr. Augustus B. Kinzel, founding president of the National Academy of Engineering, wrote in *Science:* "We will lick the problem of aging completely, so that accidents will be essentially the only cause of death." Stephen Leacock once wrote a gentle satire about a society which had reached this Utopian state—and where everybody spent all his time being terribly careful not to have an accident.

Except for the few persistent Fountain of Youth seekers who have turned up periodically in history, aging has been assumed to be inevitable, and its mechanism unfathomable. This view has been somewhat shaken in the twentieth century by a number of experiments which have demonstrated that animals, in their early stages, can be made to "mark time" in their development—by controlled underfeeding, for instance—and then resume normal growth. Dr. Clive M. McCay of Cornell was the pioneer in this field. In some of his experiments he kept young rats marking time, arresting their development, yet keeping them active and healthy, for some three years—their normal lifespan. A control group of rats whose diet was unrestricted, died as expected at the age of three. At that point the test group, subjected to no further tampering, went on to live out their full three years of normal development. McCay thus doubled their lifespan. Men have seen no way to apply this knowledge to human beings as yet; in fact, it might be dangerous to do so, since human development is vastly different from the rat's. But McCay did beautifully show that there was nothing immutable about a fixed lifespan.

The major mysteries remain, of course. What makes one man age faster than another? What makes one part of the body degenerate while another remains sound? Does each man have a built-in time clock that dictates, with no recourse or appeal, when the body has run its course? Does the organism as a whole deteriorate and die—or do individual cells wither and die; and are senility and death therefore statistical reflections of what takes place in quantities of cells? Is aging, and the rate of aging, preordained by our inheritance, or can we substantially affect it by our environment and our way of life? We do, of course, affect it, in that we can easily shorten our lives by recklessness and stupidity; but can we lengthen it beyond its normal span by behaving wisely, if we knew where wisdom lay? Dr. Comfort believes that we cannot, really, prolong our lives much beyond the present span, at least not by any techniques or medical knowledge presently available to geriatrics. Where, then, if anywhere, are the radical new techniques to come from? From radical new insights—presumably to be discovered by the gerontologists—into the nature of the aging process itself. The most fruitful of these insights may well emerge from studies of the genetic material, the cell's nucleic acids, DNA and RNA. The DNA in the original fertilized egg cell contains, like a gigantic computer, the entire set of genetic instructions that preprograms the course of the organism's life. As the original cell divides, and the new ones in turn divide, each cell gets a full quota of genetic material.* Thus the nucleus of every cell in the completed organism still has within it the entire manual of instructions. Through a series of inhibiting and triggering substances (some of the inhibitors are *histones* and some of the triggers *hormones*), the proper genes are switched on to transmit their essential instructions at the proper time, and switched off when their task is done. Some of this switching is believed to go on in the body of the cell rather than in the nucleus itself. An inhibiting chemical might, for instance, wrap itself around an enzyme to keep it from carrying out its task beyond the appointed time. Though there is a constant feedback of information between nucleus and cell body, the cell's knowhow is assumed to have been originally imparted by the DNA's genetic instructions.

Most of the nine or ten trillion cells in the average human body have the capacity to replace themselves as they wear out or die. But not all of them can do this. Among those that

---

* To clear up the confusion of those who have not kept up to date in genetics and want to know what ever happened to the chromosomes—those threadlike masses in the cell's nucleus: The chromosomes are still there, and they still make up the genetic material. The genes are parts of chromosomes, and the genes are made up of nucleic acids (DNA in higher organisms, but sometimes RNA in viruses).

35

cannot are such crucial tissues as nerves and heart muscles. One thing that surely contributes to aging, then, is the loss of irreplaceable cells. Even those that do replace themselves seem to decline in their efficiency and capability as the years go by—or at least enough of them do to affect the state of the whole organism, or vital parts of it. Cells may lose their ability to retain the same quantities of fluid; hence we get dryer and stringier as we age. Or they may become less able to absorb nutrients and eliminate waste matter, thus becoming glutted and sluggish. Many unreplaced cells are believed to turn into connective tissue; and the body's connective tissue gradually becomes fibrous, more and more resembling vulcanized rubber in texture. A significant factor in aging is believed to be damage to, or deterioration of, the genetic material itself, so that the program of instructions becomes blurred, and is transmitted with faults and errors. Dr. Comfort compares this to a photocopying process. "If we make a negative by photographing a photograph, print that negative and photograph the new photograph, and so on, the successive copies will be of lower and lower quality. It may be that a similar effect operates to make successive generations of new cells less and less effective, so that the new cells produced by an old man are in some way less viable than the new cells produced by him when he was a child."

The late Dr. Leo Szilard, after switching from nuclear physics to biology, theorized that this effect could come about as a result of genetic mutations, caused in one cell at a time or a few at a time, at random—perhaps when a cosmic ray chances to go through the nucleus, rendering the stricken cell defective or faulty in transmitting the necessary information for the perfect duplication of healthy new cells. As time goes by, more and more cells are hit in this manner until the accumulation adds up to aging, senility, and death.

Other theories compete with these, and not enough data is in to confirm the unique validity of any. But, whatever the mechanism of cell aging, an increasing number of scientists are willing to express their optimism that man will ultimately be able to control the mechanisms for his benefit—and one of these benefits ought to be a longer and more vigorous life. The essential hopeful fact is that each cell does contain *all* its original instructions—at least, up to the point where they are somehow blurred or damaged. With the prospect of manipulating the genetic material—a prospect considered overrated by some scientists, among them Dr. Barry Commoner, head of biology at Washington University in St. Louis—all kinds of possibilities loom. The cells that have lost the capacity to reproduce themselves can perhaps be taught how to do it again. Those which have deteriorated in any fashion that is susceptible to scientific unraveling can have their faulty genes corrected or replaced. One conceivable way to achieve such corrections and replacements would be by means of artificial viruses. We know that the infective cores of viruses are made of DNA or RNA. They invade the cell nucleus and either take over the cell's genetic apparatus or incorporate themselves quietly into the cell's own nucleic acids. Artificial viruses have already been made in the laboratory—the RNA type by Dr. Sol Spiegelman at the University of Illinois, and DNA viruses by Dr. Arthur Kornberg at Stanford. If viruses could be made of a man's own perfectly copied DNA molecules, then renewed amounts of it might be supplied from time to time—perhaps by injection—to revitalize the cells' failing powers.

If we learn the exact contents, structures and mechanisms of the histones and hormones—or whatever the inhibiting and triggering substances turn out to be that switch the genes on and off—then obviously such knowledge might provide the power to hold back or even somewhat reverse the course of aging. To help study aging in cells, scientists have at their command an arsenal of drugs which inhibit or encourage the cross-linking of large molecules,* and which inhibit or encourage the synthesis of DNA, RNA, and proteins. One set of experiments has provided tantalizing but inconclusive evidence that the decline of RNA synthesis (a process directed by DNA) in the liver of an aging rat could be stepped up again by the addition of fresh DNA from a young rat.

Whatever control is attained over the nucleic acid may not be enough in itself. Rehabilitative measures may have to be taken on the cell body, and on the connective tissue, and some control attained over the complex feedback processes between the nucleus and the body of the cell. But the important fact is that precise knowledge is accruing, and meaningful experiments are being proposed—and performed—at an accelerating rate.

Long before all the necessary knowledge is in for the control of cellular aging, we may chance into interim measures which could help slow down aging appreciably. As one such possibility, Dr. Denham Harman of the University of Nebraska—a chemist and physician who earlier, at the University of California, had studied the effects of radiation damage on cells—has developed an interesting theory. The effects of radiation damage, he observed, looked very much like the effects of aging. He knew that radiation produced "free radicals"—which might loosely be described as pieces of chemical compounds in search of a compound—and reasoned that the free radicals might be responsible for the effects he had observed. He knew, too, that cells normally produce some free radicals, though not in the same quantity. Could they be a contributing factor in aging? In the course of seeking to combat, repair, or minimize radiation damage, some fairly effective drugs had

---

* This cross-linkage phenomenon—the literal entanglement of long-chain molecules with one another through inadvertent chemical bonding—has been studied for many years by Dr. Johan Bjorksten, head of his own research foundation in Madison, Wisconsin. Dr. Bjorksten believes cross-linkage to be the key factor in aging.

been discovered. Harman decided to try these antiradiation drugs as possible anti-aging agents. At the end of a fifteen-month experimental period on mice which had not been irradiated, he was convinced that he had extended the animals' lives by some 25 percent. Not solid proof of anything, to be sure, but a fascinating lead that he continues to follow with increasingly encouraging results.

The hope for a real quantum-jump extension of the human lifespan, or of a major injection of new vigor into the now-senile years, seems to lie in genetic control. Implicit in such control are others—the control of cancer and the degenerative diseases. An entire organ—a liver or kidney, say, might be grown from a single cell, properly instructed and nourished; even an entire human being might theoretically be regrown from a single cell, taken from anywhere in the body, if the genetic material were in perfect condition and all triggering and inhibiting substances were placed as in the original fertilized ovum.

The estimates of experts vary quite considerably as to how long it might take before this kind of genetic engineering becomes possible. Meanwhile, doctors continue to make the most of what is at hand. Dr. Hans Selye, in *The Stress of Life,* remarks that no one really dies of old age. "We invariably die," he says, "because one vital part has worn out too early in proportion to the rest of the body. Life, the biologic chain that holds our parts together, is only as strong as its weakest vital link."

In this respect, the human organism resembles an automobile. Most cars, unlike my old Chrysler that went all at once, tend to give way one part at a time. In cars the failing parts can be repaired or replaced—an advantage they have over the human body; but perhaps not for long now.

The surgical replacement of our diseased, damaged or worn-out body parts and vital organs with artificial or transplanted substitutes could easily become commonplace in our lifetime. Transplant operations are already far from rare, though they are still performed only when all other remedies have failed. Kidney transplants have been more common than any other; somewhere between 1,500 and 2,000 have now been performed throughout the world since the first pioneering operations at the Peter Bent Brigham Hospital in Boston, and hundreds of those patients are still alive—a few for more than five years. Liver, lung, and heart transplants, not to mention a pancreas with duodenum attached, have been achieved from animal to animal, from animal to man, from man to animal, from human cadaver to man, and from man to man; that is, they have been achieved surgically, though few of the patients have lived very long. The great barrier to successful organ transplantation, as Nobel biologist Peter Brian Medawar revealed at about the same time the first kidney transplants were being carried out, is the body's rejection of the transplant as a foreign substance.* The presence of the alien tissue stimulates the body's immunological defenses to attack it as if it were an army of invading microorganisms. The transplanted organ, after functioning briefly in the body, is thus rendered useless and begins to die. In rare instances, the situation is reversed: It is the transplanted organ that rejects the body into which it has

---

* An exception to this is the cornea, the transparent front of the eye, which can be successfully transplanted from one person's eye to another's. The cornea is not rejected because (1) it is uniquely situated in a spot where it can get its blood supply without being hooked directly into the bloodstream, and thus remains inaccessible to antibodies, and (2) the very nature of the corneal cells is such that they are less likely to be affected by antibodies; therefore, they are slower to be rejected than most other body tissues. When Dr. Hector Castellanos of Temple University and Dr. S. H. Sturgis of Harvard implanted slivers of ovarian tissue in the wombs of women patients, he surrounded them with corneas to keep the tissue from being rejected. Four years later the implants were still secreting female hormones.

been implanted—in which case it is the *body,* afflicted with "runt disease," that wastes away and dies. In matching up organ recipients with potential organ donors, the typing of tissues (somewhat analogous to the typing of blood) to avoid gross incompatibilities has become standard practice; ideally, the donor and recipient should be identical twins, but few of us meet this requirement. A variety of methods—such as massive doses of radiation, potent drugs, and the draining-off of lymphatic fluid via surgical tubes—have been used to suppress the body's immune response, allowing foreign grafts to "take." But these harsh techniques also can render the patient virtually defenseless to any infection that happens to be around. When Dr. Christiaan N. Barnard, the celebrated South African surgeon, put a transplanted heart into Louis Washkansky, the operation was a complete success surgically, and the new heart kept doing its job nicely, but there was nothing the doctors could do about the pneumonia that killed the patient.

Doctors are learning how to make more effective use of immunosuppressive techniques all the time, but a much better long-range answer to the transplant barrier lies in the possibility that the immune system can be suppressed selectively. On the day when that portion of the body's defense system which rejects a foreign graft can be knocked out while leaving its disease-fighting mechanism intact, transplantation will become a much more popular operation in every major hospital in the world.

It is only a matter of time before artificial organs, too, become readily available. Heart parts, sizeable segments of blood vessels, and cardiac pacemakers are already routinely implanted, and people can live, after a fashion, on heart-lung machines and on "artificial kidneys" hooked up to the body. (These are the kinds of machines which, along with respirators, stimulators, and related equipment, can keep people going

41

on a marginal basis for days, weeks, or months after they would, in earlier days, have been pronounced dead.) When the necessary engineering breakthroughs have been made in materials, miniaturization, and internal power supplies, the day of the totally implantable vital organ will have arrived.

Without descending into complete absurdity, then, we can envision any one of ourselves who is not already too close to retirement age, walking around one day with, say, a plastic cornea; a few metal bones and Dacron arteries; with donated glands, kidney, and liver from some other person, from an organ bank, or even from an assembly line; with an artificial heart, a transformed circulatory system, and computerized electronic devices to substitute for muscular, neural, or metabolic functions that may have gone wrong. It has even been suggested—though it will almost certainly not happen in our lifetime—that brains, too, might be replaceable, either by a brain implanted from someone else, or an electronic or mechanical one of some sort. This, however, involves much more complicated surgery; Dr. Robert J. White of Western Reserve University in Cleveland, famous for having kept a monkey brain alive outside its body, believes it would be much easier to transplant an entire head*—much as the Russians have done with dogs.

It has been further speculated that, as the replacement of old organs and parts with new ones becomes routine medical procedure, transplant operations may be performed as a prophylactic measure—*before* an organ or part becomes utterly use-

---

* The late Dr. J. B. S. Haldane, always intrigued by the bizarre, wrote in 1949: "If King Charles I's or King Louis XVI's head had been stuck within a minute or so [after their executions] on a pump which supplied oxygenated blood to it, it would almost certainly have come around, after half an hour or so, enough to open its eyes and move its lips, and would probably have recovered consciousness. I hope that if I have an inoperable cancer this experiment will be tried on me."

42

less, before it begins adversely affecting other organs. Nor can the possibility be ruled out that new organs may be devised which would equip us with capacities we do not now possess.

Perhaps even more remotely in the future is the possibility that the body may be taught to regenerate its own organs and limbs, as some lower orders of animals seem capable of doing.* Research along these lines is proceeding, though not at the same pace or with the same optimism as in, say, molecular biology.

Far short of these futuristic hopes, doctors have been perfecting numerous procedures which, though much closer in their nature to traditional medical and surgical techniques, have far-reaching implications. Not the least interesting of these is one of the simplest and most direct: restarting a stopped heart. In recent years it has become commonplace to refuse to accept the cessation of the heartbeat as necessarily equivalent to the cessation of life. "I remember," says Dr. William P. Williamson of the University of Kansas, in the *Journal of the AMA,* "when cessation of heartbeat was an observation on which we simply pronounced the patient dead; now, this is a medical syndrome known as cardiac arrest"—a condition to be treated.**

---

* Salamanders, for example, normally regenerate limbs that have been amputated. At Western Reserve University in Cleveland, Dr. Marcus Singer has been able to induce frogs—which normally do not have this capacity—to regenerate severed limbs. Similar feats have been achieved by Dr. Lev V. Polezhaev in the Soviet Union, who has also claimed success in regenerating cranial bones and teeth, and expresses confidence that the heart may be induced to regenerate cardiac muscles lost through disease.

** "Cessation of respiration," says Dr. Williamson, "is a symptom also formerly implying death, which can now be corrected by an ingenious and devilishly efficient machine known as a mechanical respirator."

Though cardiac resuscitation takes place most often in hospitals, doctors have been known to stop on the street, borrow a penknife, cut open a chest, reach in and massage the heart back to life. (Few doctors would care to risk this, however; they leave themselves wide open to possible malpractice suits.) In Boston in 1967, two children in the same family "died" on successive days and were resuscitated at St. Elizabeth's Hospital. In Rio de Janiero in 1965, a pregnant woman's heart stopped beating; while her chest was being opened up, her baby was delivered by Caesarian section; by the time her heart started beating again, so had the newly-independent heart of her child. People have been revived from "death" by exposure to extreme cold, auto accidents, falls from heights, and electrocution. Most of the heart stoppages occur in the course of surgery or in the postoperative period, and many hospitals now routinely maintain emergency warning systems, standby crews, and resuscitation equipment ready to spring into action as soon as the alarm sounds to signal a stopped heart. It is not unusual for the same patient to be resuscitated many times. Nor is it any longer necessary to open the chest. Many techniques for closed-chest massage have been successfully worked out; and there now exist machines, more consistent and tireless than human hands, to do the job, with the help, if needed, of potent drugs and electrical stimulators. None of these treatments is gentle; none is without risk—of rib fracture, for instance, or organ damage. But the alternative is death—already arrived at, by any former standards.

Every now and then an unusually visionary fellow who is engaged in a BSP exercise will project his imagination forward to a time when he assumes that most of the spectacular advances which will give man control over the biology of life and death will have become reality. Based on this assumption, he calmly puts forth a proposal and a program which seem sober and sensible enough to him, but which lead a lot of sober and sensible people to look upon him as some kind of nut. But if, despite detractors, he sticks to his guns, he can attract a sizeable following. Then, the mere fact that he has attracted this kind of interest becomes in itself a fact that demands attention.

Such a visionary type is Robert C. W. Ettinger, a quiet, scholarly professor of physics at Highland Park College in Michigan. His unorthodox proposals—in which he seems to be crying out, with John Donne: "Death, thou shalt die!"—are spelled out in detail in a book called *The Prospect of Immortality*. His stubborn eloquence has inspired a rapidly growing movement in a most unlikely cause: to quick-freeze, with minimal damage, dead bodies rather than bury or cremate them. He believes it makes excellent sense to start—right now—keeping fresh corpses in cold storage against the day when they can be revived and made whole again by the medical art and science of the future. Needless to say, Ettinger's proposals have been highly controversial, and most scientists refuse to take them seriously.* Though research in whole-body freezing is being done, it seems a long way indeed from being achieved because considerable damage results from the freezing process itself—a shortcoming which Ettinger is the first to acknowledge.

The marvel is that Ettinger's notions do have some valid basis in the science of cryobiology, the freezing of living organisms, and they have struck an instantaneous public nerve. Organizations with names like the Life Extension Society and the Cryonics Society** have sprung up, their central purpose being to evangelize in Ettinger's cause. Branches already exist,

---

* He has, however, been able to assemble a reputable scientific advisory committee whose members have agreed to offer their technical advice and assistance where possible, and who have gone on record as being at least "unopposed" to his program.

** Several of these are now affiliated in the Cryonics Societies of America.

complete with officers and a zealous membership, in a number of cities in the United States and abroad. They have been working hard, and not altogether fruitlessly, to get support for their "freezer program" from doctors, lawyers, educators, clergymen, politicians, and the general public. Many believers in The Ettinger Way have made provisions in their wills to have themselves and their loved ones quick-frozen and stored as a long-shot gamble on future rescue from death. Commercial firms have designed insulated noncaskets for the purpose, and a few nonfunerals have now taken place. The first (the first to be publicized, at any rate) took place on January 12, 1966, when the body of Dr. James H. Bedford, a professor of psychology, was frozen by a team of doctors in Glendale, California, and shipped in a sealed cryocapsule to Phoenix, Arizona, where it is stored at a temperature of 320° F below zero by means of liquid nitrogen. Others have been frozen since, and at least seven hundred more have registered their desire to be frozen with the Life Extension Society in Washington. Some members of the Life Extension Society carry membership cards and Medic Alert wristbands with emergency freezing instructions telling how to proceed in case they should die; and some chapters of the Cryonics Societies have fitted out mobile rescue units equipped for fast freezing.

Even if Ettinger's ideas turn out to be totally unsound scientifically, the mere fact that large numbers of people *believe* them to be sound, and act upon them, will raise the same kind of problems—including scientific and biomedical problems—as if they were in fact sound.

"We have heard of such abstractions as a germ-free world, indefinite lifespan, and the intelligent self-reproducing machine," said Dr. Lederberg, in the discussion period following a seminar on futuristic biology. "Each of these is quite possibly not attainable in its full form." The same may turn out to be true for the synthesis of life in the laboratory, or man's total mastery of the genetic material. "But it doesn't need to be so," Lederberg believes, "in order to be well worth thinking about. These abstractions pose problems that we have to deal with, either in emulating life or in setting up appropriate social dynamics in the clearest possible form. There is no point in arguing whether we will fully understand the system. We may never fully understand any mechanical system, and yet it is of the utmost value to postulate one that is frictionless in order to isolate other elements of it."

We can look upon the freezer society as an abstraction, like the probably-never-to-exist frictionless machine, as a means of highlighting the many and varied quandaries which BSP evokes. And it is time that we began to take a closer look at some of these quandaries.

It should be clear by now that in the practice of BSP there can be no omniscience. No expert is qualified to tell us which, among the possibilities so far described, will be achievable. But suppose *all* of them turn out to be things we can do. Should we do them just because we can? Do not some of the prospects begin to look spooky?

Not everyone, certainly, is ready to holler Hurrah for all the millennial contingencies inherent in man's coming control of life. Their mere evocation usually elicits as much dismay as delight. But to hold up the research that leads in those directions because of our qualms, would be to delay indefinitely the attainment of the immediate medical benefits. Since hardly any of us want to forego these benefits—at least, insofar as they apply to each of us personally, or to our loved ones—it appears certain that research and experimentation will continue to proceed apace.

While the advantages we all hope to gain in terms of better therapies and longer lives are relatively noncontroversial, the road to these gains—the research and experimentation itself as well as the timing and manner in which the results are applied—will certainly be controversial all the way. Already familiar subjects of debate are the moral, ethical, and legal dilemmas posed by transplant surgery.

Does anyone have the right to ask a man to give up his spare kidney and undergo the major surgery that kidney removal entails, in order to give another man—even a close relative—a chance, and not always a good chance, to survive for a limited extra time? Transplant surgeons are haunted by the possibility that they might lose a donor on the operating table. In that case, they would have been instrumental in killing a healthy man on the long gamble of saving a sick one. "Every physician recognizes the increased risk of death," said Dr. J. Russell Elkinton in a 1964 editorial in *Annals of Internal Medicine*, "in a patient with only one kidney who develops pyelonephritis or throws an embolus to his renal artery. Is it right to subject a healthy person to this risk, to the possibility (though not the probability) of shortening his life by 25 or 30 years in order to extend another's life by 25 or 30 months or less?"

"Even more serious," Elkinton believes, "is the possible disruption of the mental and emotional health of the potential donor and of the patient's family. I know of a brother of a patient who declined to donate his kidney—with resultant severe emotional trauma; I know of another family torn apart by a mother giving a kidney to her child against the wishes of the husband and father. Such psychosocial complications occur in many difficult clinical situations but never more so than in this one of organ transplantation."*

---

* In a recent sermon, Canon Michael Hamilton of the Washington Cathedral said, "As a Christian, I believe that a man with family responsibilities is not clearly bound to endanger his own health by, let us say, donating a kidney to his brother."

It is not even clear that a person has a legal right to volunteer one of his organs (especially if he is a minor), to authorize the posthumous removal of an organ from a dying relative, or even to will his own organs for use after death. Many of the laws governing cadavers are based on precedents in English common law that go back to the sixteenth century. These precedents hold, for the most part, that, after a man's death, the next of kin inherits no property rights that would permit him to dispose of any of the corpse; he gains custody of it for burial only.

Some states, under medical prodding, are trying to rectify these legal shortcomings by permitting anyone to will his organs for transplantation—or, if he has neglected to do so, to permit his next of kin to give consent.* Some proposed laws would allow doctors to remove organs from bodies with or without consent—or even to permit qualified coroners to remove them in the course of routine autopsies. At the moment, though, there is no way to guarantee absolute legal safety in these experimental areas, a situation many doctors find frustrating. Society, says Dr. Henry K. Beecher of Harvard, "can ill afford to discard the tissues and organs of hopelessly unconscious patients so greatly needed for study and experimental trial to help those who can be salvaged."

Dr. Roy Cohn, professor of surgery at Stanford, tells of an episode where a young man had been accidentally shot in the head, effectively destroying his brain but leaving his organs

---

* In September of 1967, Massachusetts became the first state to put into effect a law "facilitating anatomical donations," a law worked out by Dr. William J. Curran, a Boston University law professor, in consultation with surgeons in the Boston area. In July 1968, the National Conference of Commissioners on Uniform State Laws approved a proposed new "Uniform Anatomical Gift Act" in the hope that all 50 states would eventually adopt it.

49

intact while he was kept alive by an artificial respirator. "The man's family offered to donate his kidneys, Cohn recalls, "but the uncertainties of the law regarding the right of the patient's family to make the offer, as well as the responsibilities of the physicians in accepting this patient, were so indefinite that those priceless kidneys could not be used."

Not all surgeons play it so safe—and a few have been sorry later that they did not. There was one case where a man who had been beaten to death was brought into a hospital. His wife gave permission to remove one of his kidneys for transplant to a dying patient. For twenty-four hours the doctors used a heart-lung machine to keep up an artificial circulation and heartbeat, to be certain the kidney would remain fresh and undamaged. The victim's assailant, when he came to trial, insisted it was not he who had killed the man—who was, after all, still alive in the hospital for a time. It was the doctors who had done the poor man in, he argued, by removing his kidney and turning off the machine. (His arguments were not upheld by the court.)

In more than one instance transplanted kidneys, unbeknownst to the doctors, have brought with them active cancer cells. In these cases, the transplants were a success, but the patients died of cancer.

Apart from the moral predicaments posed for the medical profession and for society at large, there are challenges—on a more personal, philosophical level—to some of our oldest and most cherished concepts about ourselves. A man could always take for granted, for instance, the premise that he was he, himself, a unique specific person (I, John Doe), an individual with a name and a definite and recognizable identity which he would retain through whatever vicissitudes for the duration of his lifetime, from birth to death. And he knew that when his life came to an end—well, that was the end; he was dead. A man was alive, or he was dead—irrevocably and forever—and there was no in-between nonsense. But these formerly clearcut concepts of identity and death have been rendered ambiguous by our recent biomedical tamperings.

Writers and artists already do a lot of articulate worrying about how alienated they feel, and psychiatrists' couches are constantly occupied by people who say, in effect: "Doctor, I don't know who I am." So, even before the Age of Biology has set in, modern man faces a crisis of identity. Organ transplants and artificial implants will probably aggravate this crisis.

"Cannibalizing," says Dr. Elkinton, "was the term applied to a practice that unhappy circumstances spawned in some of the more remote areas of action in World War II. This process consisted in combining parts of a number of damaged vehicles to make one whole vehicle that would function." He suggests that we are learning similarly to "shore up some damaged human 'vehicle' and put him on the road again. . . . The analogy may be a bit stretched," he admits, "but the possibility of such human cannibalizing is implicit in the development of artificial internal organs and in the experimental transplantation of one human being to another." Almost any one of us might become, through surgery, such a cannibalized being.

The term most surgeons have come to prefer for such a being is a "chimera." In Greek mythology, the Chimera was a composite monster with a dragon's tail, a goat's body, and a lion's head that spit fire. "Curiously enough," Dr. Francis D. Moore points out in his excellent primer on transplantation, *Give and Take,* "the Greek root for this connotes the concept of a very young goat." Thus the term seems even more appropriate as applied to the results of techniques designed to frustrate the ravages of aging and hold off the approach of death. "Even in mythology," says Dr. Moore, "youthfulness

51

had something to do with chimerism." Now "the term chimera has been lifted bodily from mythology to immunology to signify *any organism carrying within it two or more healthy living tissues of different genetic origins."* (The italics are his.)

"What," asked Dr. Lederberg, "is the moral, legal or psychiatric identity of an artificial chimera?" How many of a person's parts can be replaced, and how considerably can his character and outlook be altered in the process, before he actually becomes a new and different person? Dr. Seymour Kety, an outstanding psychiatric authority who recently moved to Harvard from the National Institute of Mental Health, points out that fairly radical personality changes have already been wrought by existing techniques such as brainwashing, electroshock therapy, and prefrontal lobotomy, without raising serious questions of identity. But would it be the same if alien parts and substances were substituted for the person's own, resulting in a new biochemistry and a new personality with new tastes, new talents, new political views—perhaps even a different memory of different experiences? Might such a man's wife decide she no longer recognized him as her husband and that he was, in fact, not? Or might he decide that his old home, job, and family were not to his liking and feel free to chuck the whole setup that may have been quite congenial to the old person? It is hard to say which would be more disquieting—ambiguity about one's own identity, or uncertainty about the identity of others.

Brain transplants, if they ever became feasible, would certainly accentuate the dilemma. Dr. Moore, who is a pioneer transplant surgeon of Harvard and the Peter Bent Brigham Hospital, believes that the idea of transplanting whole brains from one person to another may turn out to be "nothing but idle chatter." But he adds that "if it were accomplished, the recipient would be a different person." Brain surgeon Robert J. White feels the same way. A new brain—or a new head-with-brain, most startling of chimeras so far suggested—would entirely change the recipient's identity.*

Not that acute problems of identity need await the day when the wholesale replacement of vital organs is a reality. Very small changes in the brain might well bring about astounding metamorphoses in behavior and attitude. Scientists who specialize in the electrical probing of the brain have, in recent years, been exploring a small segment of the brain's limbic system called the amygdala—and discovering that it is the seat of many of our basic passions and drives, including the drives that lead to uncontrollable sexual extremes such as satyriasis and nymphomania. Subtle changes in the chemistry or a slight scrambling in the electrical circuitry in areas of the amygdala in close proximity to one another might well determine whether a person is terrible-tempered or meek, whether he is the victim of unregulated sexual urges, or a person scarcely interested in sex at all.

Suppose, at a time that may be surprisingly near at hand, the police were to trap a sadistic rapist whose crimes had terrorized the women of a neighborhood for months. Suppose further that, after he has been through all the current legal procedures that apply to his circumstances, he is sent—not off to jail, but into the hospital for brain surgery. The surgeon delicately readjusts the distorted amygdala, and the criminal has now become a gentle soul with a sweet, loving disposition. He would not do violence to a flea. He is shocked to hear of the

---

* Hardly to be touched on in this book is the practice of transsexual surgery, which has begun to gain respectability in some of our major medical centers. This involves truly radical surgery, including—in the case of a man wanting to become a woman—the removal of all his sex organs and the creation of artificial female ones, followed by extensive hormone therapy. When a surgeon can look at a stripteaser performing in a night club and say, "I remember her when she was a quiet young fellow," then change of identity has ceased to be academic.

kind of crimes his former self had committed. The man is clearly a stranger to the man who was wheeled into the operating room. *Is* he the same man, really? Is he responsible for the crimes that he—or that other person—committed? Can he be punished? Should he go free?

As time goes on, it may be personally desirable and socially necessary to declare—without the occurrence of death—that Mr. X has ceased to exist, and that Mr. Y has begun to be. This would be a metaphorical kind of death and rebirth, but quite real psychologically—and perhaps legally. Mr. X might have to be declared dead, and a birth certificate issued for Mr. Y. (What would be listed as the cause of death? What age would be recorded on the birth—or rebirth—certificate?)

But even death in the old sense, the physical death of the body, will be harder and harder to pin down with any precision. Medical history is replete with spectacular life-prolonging, death-postponing feats—but in the past none of these have seriously brought into question the basic concept that there was a sharp line of demarcation between life and death. There was an agreed-upon moment when breathing stopped, when the heart stood still, when the vital organs quit working, and the doctor could solemnly pronounce a man dead. How many mystery stories have you read, how many TV dramas have you watched, with a doctor testifying authoritatively from the witness stand that death occurred precisely at 2:28 A.M.? Now that it has become commonplace to resuscitate the "dead," the final pronouncement is not so easily given—and few doctors would be so fatuous as to pinpoint the moment in time.*

No wonder, then, that medical journals these days so fre-

---

* There is a case on record of a man who, after being worked over for hours at a Midwestern hospital, was finally given up as beyond retrieval. His wife, notified of his death by phone, collapsed and died of a heart attack. Meanwhile, the man on the table astounded everyone by starting to breathe again—and he survived his widow.

quently carry articles with titles like, "When Is a Patient Officially Dead?" Doctors have lately been meticulously distinguishing among several categories of death. They differentiate, for example, between clinical death (from which a patient may now hope to be revived—not only once, but perhaps many times) and biological death, in which the cells and organs, and especially the brain, are irreversibly damaged. (Robert Ettinger, in his campaign for the freezer society, insists that in the future practically *no* damage will be irreversible.)

As previously described instances already amply demonstrate, a precise and universally acceptable definition of death becomes critically important for surgeons interested in transplant operations. If the Stanford team had been able to declare dead the young man whose brain had been shot away, despite his continued artificial heartbeat, they could have used his healthy kidneys for transplantation into two other people dying of kidney failure. So three people had to die, where two of them might have been saved—at least, temporarily. (In the case of the first South African heart transplant, the doctors *were* willing to consider the young lady dead on the basis of brain destruction in an automobile accident. When her heart stopped, they did not try to restart it—until it was in the recipient's chest.)

"I have seen patients," says Dr. Williamson, of the University of Kansas, "with brain-stem failure, with dilated fixed pupils, decerebrate rigidity, and cessation of spontaneous respiration, who have had a tracheostomy and were assisted with a mechanical respirator. With fluids, electrolytes, and good nursing care, the essentially isolated heart in such a patient can sometimes be kept beating for a week." Is such a person alive? Would it be murder to turn off the machine, let the heart stop, then remove it for transplant?

Such considerations are obviously difficult to resolve in the case of the heart. A heart is not only a unique and indispensable organ—but it is also quite perishable. Where a kidney allows roughly three hours of leeway between death and transplantation, a heart must be removed within thirty minutes after cessation if it is to remain suitably fresh. The spate of heart transplants which have been performed since Dr. Barnard led the way have demonstrated convincingly that there is no surgical barrier to such transplantation—which, incidentally, had been done hundreds of times in dogs. The big barrier to heart transplants, apart from the immunological barrier, would be the scarcity of circumstances under which donor hearts could be procured.

Before discussing these procurement problems, it is well to point out that the selection of ideal *recipients* is far from easy. An outstanding heart surgeon told me privately that he would hesitate ever to take out a man's heart before death because even a diseased heart has remarkable and unpredictable recuperative powers. At least one patient with heart failure thought to be irreversible was being seriously considered as a transplant prospect when he fooled everybody and got well enough to go back home. The fantastic resilience of even a badly damaged heart was nowhere more strikingly demonstrated than at the Wilhelmina Hospital of the University of Amsterdam. A seventy-year-old man had died of a massive heart attack. An hour and a half later—an hour beyond the time considered safe for transplanting—a team of surgeons removed the diseased, deteriorated, seventy-year-old heart from the corpse. They hooked it up to a machine designed to flow oxygenated ox blood through the coronary blood vessels. Then they opened the stopcock to start the flow. "It was astonishing," said Dr. Dirk Durer, one of the surgeons, reporting the incident later at a medical meeting. Without the necessity for any mechanical stimulation or electric shock, the "dead" heart started beating as soon as the ox blood began to flow. It continued to beat without any further help for six hours. The experiment was terminated, not because the heart stopped at that point, but because the researchers were too tired to go on.

Despite the heart's surprising comeback capabilities, there are times when doctors are virtually unanimous on the hopelessness of a given case, and are therefore willing to try implanting a new heart as a last-chance possibility of saving his life. According to Dr. George Reed, New York University heart surgeon, these would be cases where there had been multiple heart attacks, the coronary arteries had been damaged beyond repair, there was great destruction of heart-muscle fiber and the patient was in a chronic state of heart failure—all these things; or where the patient was afflicted by one of the little-understood conditions known as "myopathies" in which, without heart attacks, the cardiac muscles have simply deteriorated to the point where the heart can no longer do its pumping job.* A mere heart attack, even a very bad one, does not justify a transplant attempt; too often, the patient makes a complete recovery—as witness Presidents Eisenhower and Johnson. Moreover, many chronic heart conditions, formerly fatal, can now be corrected by a variety of surgical techniques which are less drastic, and have more hope of success, than the transplant. So the percentage of heart patients who might be candidates for heart transplants is, in Dr. Reed's opinion, very small. Doctors' personal feelings on this were aired in a poll conducted by the Marion Laboratories of Kansas City early in 1968. Of 218 cardiologists questioned, more than half said they would not consent to heart transplants for themselves even "if they had advanced heart disease with a poor prognosis."

If the choice of recipients were no problem, getting donor

---

* On the day medicine solves the basic immunological problems and heart transplantation becomes routine, then the preferred recipients may become very young people who happen to have heart trouble but are otherwise reasonably healthy.

hearts still would be. "Even after the ultimate immunological barrier has been overcome," Dr. Michael E. DeBakey, Houston's famed heart surgeon, has pointed out, "it will not open up any royal road to wholesale heart transplantation. I do not foresee that there will ever be an adequate supply of donor hearts." This conviction is one of the major reasons for De-Bakey's own concentration on the development of artificial hearts.

Dr. C. William Hall, an associate of DeBakey's in the artificial heart program, once expressed similar sentiments. "Can you think of any set of circumstances," he was asked, "where a heart capable of being restarted in another body would no longer be of any use to the body from which it was taken?" He thought a minute and said, "Yes. I can think of one set of circumstances and only one: If we still used the guillotine as a means of execution, completely separating the head from the body, then I would say that the headless body had no further use for the heart either."

There is one other circumstance, short of decapitation, which some scientists believe would justify removal of the heart—and which Dr. Hall agrees would be the functional equivalent of beheading. "The answer," says Dr. Moore, "will always be that the brain was irreversibly damaged." This was the justification used by Dr. Barnard, and by Dr. Norman Shumway of Stanford. It was also the justification for Dr. Adrian Kantrowitz of Maimonides Hospital in Brooklyn; in his first case, it was an anencephalic baby—that is, one born virtually without a brain, thus with no chance of survival. When Dr. Kantrowitz had as a patient a baby born with a heart so weak and so defective as to doom the child to certain death, he sent out a call to hundreds of hospitals all around the country, found a just-born anencephalic baby whose parents gave their consent to the operation. The idea in each case is to take two doomed individuals, and save the one that can be saved through the creation of an artificial chimera.

Irreversible brain damage was also considered an adequate justification for heart removal by Dr. James D. Hardy of the University of Mississippi, who was the first surgeon to perform a lung transplant and who, except for a freak of circumstance, might have been the first to do a heart transplant, years before Dr. Barnard and the others. On three occasions Hardy had potential donors on tap, keeping them alive artificially; but the timing was never quite right for the potential recipient. On another occasion, with the recipient ready and the surgeons not able to wait any longer, they could not with certainty pronounce the donor dead, so they proceeded to implant in the patient a heart which had been taken from a chimpanzee. The chimp heart worked for a while, but finally proved inadequate to the task of pumping a human being's blood supply.

Dr. Hannibal Hamlin of Harvard, among others, has been urging that death be defined as death of the brain, ascertainable by electroencephalograph (EEG) readings. A victim of brain death whose circulation and respiration are kept going artificially is, in Hamlin's view, merely a "heart-lung preparation." The patient as a *person* "has ceased to exist," he says, "if his technically accurate EEG has shown flat record"—i.e., no electrical activity in the brain—"for 60 minutes." Hamlin emphasizes that he would never imply that "it might be possible to identify with temporal precision such a complicated biologic instant in the dying patient."

"Although physicians cannot yet specify the exact moment of death," says Hamlin in *Medical Tribune,* "medical opinion should be able nonetheless effectively to promulgate understanding how individual survival as understood by the physician has ceased to exist." Moreover, "medical science should strive to obtain practical agreement from Church and Law on the recognition of death as it affects the vital organs that govern the total organism, notwithstanding the persistent semantic and multilingual difficulties of formulating a definition satisfactory to all concerned."

Cognizant of these difficulties, the French Academy of Medicine appointed a high-powered committee, headed by the distinguished surgeon François de Gaudart d'Allaines, to study them. After four months of pondering, the committee came up with a report, a report which the Academy unanimously adopted by a show-of-hands vote. What the Academy did, in effect, was to make a firm decision about the definition of death—a decision not too different from the one Dr. Hamlin has been urging on American doctors. Death, they agreed, would henceforth be considered to be the death of the brain, and the brain would be considered dead if negative EEG readings persisted for forty-eight hours.*

The patient could, at the end of that time, be declared officially dead by a committee of three doctors, even if he was still being kept "alive" artificially, and even if he might be kept alive in this same sense for an indefinite period of time. Acceptance of this definition would mean that organs could be removed from the officially dead body for transplantation to other patients who were in danger of dying for lack of them.

The Academy's recommendations certainly did not received anything like automatic acceptance even in France, not

---

* In August, 1968, a panel consisting of thirteen members of the Harvard faculty headed by Dr. Beecher recommended the adoption of the following criteria to define "irreversible coma": (1) Total unresponsiveness to even the sharpest stimuli; (2) total absence of any movement or breathing after an hour's careful observation; (3) total absence of reflexes; (4) a flat EEG. All tests should be rerun again after twenty-four hours to be certain no change had taken place in the interim. Even these criteria are not final in those cases where the patient's body had been chilled to lower than 90 degrees or depressed by large doses of barbiturates.

even among the medical population. When a patient is still breathing and his heart is still beating, many doctors would still feel rather uneasy about signing a death certificate, no matter how inactive the brain. Nor is any similar recommendation likely to get instant, universal recognition. "This is a very intriguing problem," said a prominent American surgeon, commenting on the French Academy's action. "It must be approached very carefully to avoid the day when someone says, 'You let my father die just to give his kidney to another person.'" In such a case doctors would indeed leave themselves open to suit or prosecution. As a matter of fact, even the full consent and approval of the next of kin is no insurance against this.

A case in point occurred in Sweden, causing a public sensation. A team of doctors headed by Dr. Stig Ekestrom at Stockholm's Karolinska Hospital transplanted a kidney from a forty-year-old woman who was dying of a cerebral hemorrhage to a forty-five-year-old man suffering from kidney failure. The woman's condition was deemed hopeless, and her husband consented to the transplant. She died in a respirator two days after surgery, and the man who got the kidney died two weeks later. Sweden's attorney general, K. E. Walberg, ordered an investigation to see if legal action was warranted. "If death was hastened by the removal of the kidney," he said, "they may have been guilty of intentional killing or causing another person's death."

The surgeons' superior, Dr. Clarence Crafoord, lent his full moral support. "If you wait until death, in the conventional sense," he argued, "the possibility of a successful transplantation is decreased. We had also agreed that the closest relative of the kidney donor had to give unqualified consent." But Dr. Lars Leksell, head of another department in the same hospital, felt otherwise. He called the whole episode "frightening." "The fact is that a patient's condition is judged hopeless," he

61

said, pointing out that this is often extremely difficult to ascertain, "cannot give the doctor the right to dispose of his organs without his consent."

Dr. John P. Merrill, Dr. Moore's eminent colleague at the Peter Bent Brigham Hospital, agreed vehemently at the time. "I think this procedure was unethical and a backward step," he said in *Medical World News*. "In this day of the pump and respirator, the criteria of death are often difficult to determine. If you take a kidney from a dying person without consent, where do you stop? Next time you might take it from a person who doesn't die, and that would not only be immoral, it would be criminal."

The same week that the fourth and fifth human heart transplants were taking place in Capetown and Brooklyn, a friend of mine, an eminent authority on cardiovascular diseases, was in Princeton attending a conference on strokes. The favorite joke at the meeting, he told me when he got back, had one doctor advising another, "If you *must* have a stroke, be sure to fall *forward*." The implication was clear: With so many surgeons eagerly seeking heart donors, it would not be prudent to fall on your back, thus exposing your chest.

Later I had occasion to stop at a friend's office, where the topic was again heart transplants. A young man said to a strapping colleague, "Hey, you better quit going around looking so healthy, Fella—it might be dangerous." This black humor was a reflection of the swift shift in attitude that had taken place in regard to heart transplants. After the first reports of Dr. Barnard's historic operation, there was a great glow of enthusiasm and a general feeling that the gift of one person's heart to save another was a marvelous new possibility. But when four other transplant operations were performed in such lightning-quick succession, some reservations began to be expressed about donating hearts.

Another friend of mine, this one not a doctor—and not

joking—said, "You know, with all this going on, I would be afraid to get badly hurt or seriously ill. Would the doctors work hard to save me, I wonder, or would they look upon me as a prospective heart donor, and wish for me to die?" A woman hearing these remarks reminded us of that obscene episode in the movie, *Zorba the Greek*—where an old lady lay dying, and the voracious neighboring villagers hovered around obsequiously waiting to loot her possessions at the instant of her death. Then, finally, too impatient to await the ultimate death rattle, the villagers ghoulishly stripped the house bare while the old lady screamed out her last breath.

Few people would put transplant surgeons in this ghoulish category. But the comparison did not seem at all farfetched to Dr. Werner Forssman, the Nobel-prizewinning German surgeon, who, after Dr. Barnard's first transplant, conjured up heart-excision tableaux reminiscent of Aztec sacrificial rites. "Is it not a gruesome scene," he asked, writing in the *Frankfurter Allgemeine Zeitung,* "if physicians in one operating room treat a patient with the heart-lung machine while in the next room a second operating team stands around a young woman with their knives ready, not to help her but glowing with greediness to exploit her defenseless body?"

Dr. Forssman's remarks were widely quoted by the wire services. Though his viewpoint was extreme, there were other doctors who felt that the prevalent qualms were anything but absurd. Neurologists and neurosurgeons have seen some remarkable recoveries from strokes and brain hemorrhages, especially among younger men and women—the types who would make the best heart donors. "If a colleague wants an organ for transplant," says Dr. Joseph Ransohoff, head of neurosurgery at the New York University Medical Center, "my department is where he comes looking for it. Since time is of the essence, he may be eager for me to pronounce the prospective donor dead. But he simply has to wait until I am absolutely certain that the person *is* dead, totally and beyond recall, even if it risks the waste of a precious organ. The responsibility is even greater in those instances where the dying person is to be transferred from the place where he is being treated to a more convenient site, adjacent to the recipient. In any such case there must be no doubt whatsoever that the patient's condition is indeed beyond hope."

Another doctor said he would refuse to transfer a patient in this manner. "When a doctor really wants an organ from a dying man," he says, "then I simply cannot have 100 percent confidence that there will be a 100 percent effort to keep him from dying."

Dr. Ransohoff goes through a complex procedure lasting some twelve hours, including the use of drugs that stimulate brainstem activity as well as blood pressure, before he is willing to declare a patient truly dead. Some doctors insist that, even after brain death, if the heartbeat and respiration are maintained artificially (which must be done if the heart or other organs are to be kept fresh enough for transplant), then the person is still technically alive. But on the whole, most surgical teams contemplating the removal of organs from a body, feel they are on safe ground if they ascertain that the brain has been electrically inactive for a sufficient period of time, and that the *natural* heartbeat and all *spontaneous* respiration have ceased. All three criteria must be satisfied to be sure that a cadaver is really a cadaver.

Visiting the United States shortly after the first heart transplant, Dr. Barnard granted an interview to a New York-based Soviet news correspondent, Geinrich Borovic, who promptly cabled a set of lurid speculations to his paper, *Komsomolskaya Pravda:* "Just imagine a bandit corporation which deals with the murder of people only for the sake of selling their organs on the black market. These goods," he explained, "would be needed by doctors for their rich clients. Money could make doctors register death before it happened. Money can make people sell their organs before their deaths."

Absurd?

Dr. Lederberg suggested years earlier that there might some day be a thriving black market in organs. As for the corruptive power of money, Dr. Irvine H. Page, recently retired as research director of the Cleveland Clinic but still editor of *Modern Medicine,* believes it should not be underestimated. "The entrance of money into the situation is a matter of real concern," he believes. "If the recipients of hearts, or their doctors, can accept large sums of money for, say, TV rights to their stories, then certainly prospective donors or their next of kin will begin to want money for the donated organs. There has already been a well-publicized instance where a disillusioned veteran offered his heart for sale for a million dollars."

Thus what began as a humanitarian effort could become a blatantly commercial enterprise. With donor hearts in such short supply, *would* the well-to-do get preference? "It is up to the medical profession to set standards all along the line," says Dr. Page, "so we can be proud, and not ashamed."

When and if all the ethical and legal ambiguities are resolved, it would seem that transplant surgeons could gain access to a hopefully major source of transplantable organs—human cadavers; and the potential supply will be considerably augmented when ways are worked out to preserve them for longer periods of time in organ banks. By then perhaps the public will have been educated to overcome its fears and scruples, and most people would be willing to give permission before death to have their organs removed after death.

But all these hopes could be torpedoed if Robert Ettinger's freezer program suddenly catches on. And considering the progress of the movement thus far, it cannot be brushed summarily aside. To those who find the whole idea of body-freezing repugnant, Ettinger says cheerfully, "To each his own. Rot in the best of health." To those who hold their heads at the horrendous aggravation of the population problem implicit in the freezer program, Ettinger replies, "No one in his right mind will accept death for himself simply because there is a possibility that, at some future date, he might crowd someone." It is not that he fails to recognize the enormity of the social and economic issues raised by his vision of immortality. He is simply convinced that, no matter what the issues, they will have to be somehow dealt with, because as far as he is concerned the freezer program is inevitable.

In the event that it comes to pass as Ettinger predicts, what then happens to our notion of death? In the past, when a man was pronounced dead, he was dead medically, dead legally, and dead theologically. For theologians and jurists, adjustment to the new status of death will be hard. For the medical profession, accustomed as it is to changing its attitudes to keep up with changing practices, it will be easier. "After all," says Ettinger, "the freezer program is just another medical measure to prolong life—hardly more bizarre than an iron lung, hardly more unnatural than penicillin, hardly more radical than a kidney transplant."

Ettinger's views got some support recently from the eminent French biologist, Dr. Jean Rostand who, after the aforementioned decision of the French Academy regarding a new definition of death, commented in *France-Soir:* "A dead man is only temporarily incurable. The notion of physiological death has changed through the years. And it is possible that a man considered dead in 1966 might not be considered dead if we were in the year 2000."

If this viewpoint gains impetus, what will it do to the surgeons' increasing hopes for using cadaver parts? Many people already bequeath their eyes, for example, to eye banks, to be used by surgeons to restore sight to the blind. The long-range plan has been to establish storage banks for all kinds of organs and body parts that could be drawn on to implant in live patients who needed them—to create a new population of cannibalized beings or chimeras. As already indicated, some surgeons advocate laws that would enable them to remove needed parts from cadavers without special permission. But if there existed a viable freezer program, cadavers could no longer be so considered, and removing organs from them for research or for healing would be stealing parts they might one day need themselves.

Where, then, would the needed organs come from? Animals, perhaps. Ape kidneys, pig livers, and a chimpanzee heart have already kept human patients alive for short periods of time. Dr. Lederberg suggested at a Ciba Foundation symposium in London that animals might be especially bred to supply genetically reliable organs for people. "Another method with possibilities but without immediate practicability," Fred Warshofsky suggests in *The Rebuilt Man,* "is the use of fetal or embryonic material from which adult-sized organs and tissues may be grown . . ."

"The potentialities of such embryonic farming," says Warshofsky, "are staggering. Imagine limbs, kidneys, hearts, lungs, livers, skin, bone, nerves, eyes, teeth, all grown to order and compatible with the recipient. For embryonic tissue apparently has no immunological activity. It cannot provoke the defense mechanism of the recipient into countermeasures. It will join the body not as a foreign antigen, but as a natural protein."

Listen to Dr. Rostand's description of some of Dr. Etienne Wolfe's experiments along these lines in France: "He removed sexual glands [of chickens] when they were still at the stage of undifferentiated suggestions, shin bones when they were tiny wands of cartilage and fragments of skin, or eyes when they were still minute translucent vesicles attached to the brain. And in the small saltcellar which served as the receptacle for the culture, he watched the sexual gland turn into an ovary or a testicle, he watched the shin bone lengthen and become curved and mould its extremities as though in articulation with a phantom thigh bone and foot, he watched the skin bristle with budding feathers as it took on its characteristic look of 'goose flesh,' and he watched the globe of the eye swell and acquire color."

If organs could be grown from embryonic tissue, ways could certainly be found to keep them alive—just as the Dutch surgeons mentioned earlier kept a human heart alive, and just as the Cleveland researchers kept a monkey brain alive.* Hopefully, by the time these developments come to pass, the storage problem will have been solved.

Another hope, also mentioned earlier, is that organs could be raised in tissue culture, perhaps even from the patients' own cells. The practice of transplantation might conceivably have to be dropped altogether in favor of miniaturized implantable artificial parts. There is another, though more remote, hope— the possibility that the body might be taught to regenerate its own parts. "We cannot," says Dr. Singer, the man who has taught frogs to regenerate severed limbs, "rule out the possibility that some day human beings might be able to regrow organs and tissues which they presently cannot."

Procuring body parts, however, will not be nearly as vexing or as immediate a problem as procuring whole bodies. Thousands of bodies are needed every year so medical students can properly study anatomy, and prospective surgeons develop their skills—especially when operating procedures are getting more daring, more delicate, more demanding all the time. At most medical institutions there has been a chronic scarcity of cadavers, a scarcity somewhat alleviated by a recent trend for

---

* Dr. I. Suda and his associates at Kobe Medical College in Japan brought the isolated brain of a cat back to a semblance of normal electrical activity after seven months in the deepfreeze.

people to will their bodies to medical schools, thus not only saving funeral expenses but also contributing to the general welfare. But who will elect to have his corpse dissected if there is hope that the intact body might one day be revived? Almost overnight, if freezing becomes fashionable, the cadaver shortage could become acute. Ingenious body models might be built, and animal studies would continue, but there is no known substitute for the direct, firsthand study of the human body in all its magnificent intricacy. The conscience of the future might find it acceptable to raise brainless, unfeeling beings in tissue culture or *in vitro,* then declare them to be not human and therefore usable for research purposes. But barring some such drastic and, to us today, disquieting measures, the freezer society might face the irony of a situation where there were no skilled surgeons or resuscitators to bring the dead to life because there were no cadavers to use for research and training.

In the freezer era, any mutilation of a corpse—its status as a corpse being after all only tentative—would be undesirable. Even standard autopsy precedures would be challenged—as they are occasionally anyway on religious grounds—and the laws governing these procedures would have to be revised. This would be among the least of the challenges posed for legislators and jurists by the partial abolition of death—or, at least, its indefinite postponement.

When a man is put into the deepfreeze instead of into the grave, what happens to his estate? Are his heirs forever denied their inheritance? Is his property to be kept in trust for him, unusable by anyone else, perhaps for centuries? Can his widow —or, rather, his wife—or his children collect his insurance? What about this man's retirement benefits, his veteran's pension? Do these go on being paid virtually in perpetuity, accumulating, with interest, to his account? And if overpopulation is already a worry, think of adding all those revivable corpses every generation.

Even without a freezer program, individual lives may be quite substantially prolonged by the gradual eradication of cancer and heart disease, by the replacement of failing body parts, and by breakthroughs which permit a real slowdown, or even reversal, of the body's aging processes. Future societies then, will almost certainly have more and more vigorous oldsters hanging around for longer and longer periods of time. For a still robust man, sixty-five is a ridiculously young age at which to go into retirement. It will look more and more ridiculous as time goes on. As the years stretch out, industrial retirement benefits and government Social Security plans will all be hopelessly inadequate. Old people will be holding on to their positions and their property for longer and longer, making it harder and harder for an increasingly impatient set of young people—in an ever more automated and perhaps overcrowded planet—to acquire positions and property of their own. If, when these old people die, they have themselves frozen, they will still want to keep it all so it will be there waiting when they awaken from their long frozen sleep.

And what about lines of succession? If a president "dies," but is not really dead, can the vice president ever become more than acting president? In the United States, the president's term does expire and another one can be elected. But there are nations where the heads of government hold lifetime tenure. Could anyone ever replace a dead-but-frozen ruler? As billions of people were frozen in every generation, the dead might ultimately own all the property and hold all the important offices. Barring the return of some sort of ancestor worship or a cult of the dead, would the unfrozen population put up with it? With life so overabundant and so easy to produce anew, why should anyone want to go to all the trouble and expense of reviving all that competition? Those piled-up frozen corpses might in fact be fiercely resented. If the planet became desperately overpopulated, with prevailing undernourishment and contempt for life,

70

their very presence might even encourage a drift toward canni-
balism. Of course, a few gigantic power failures, like the one
that recently hit the Eastern seaboard, and the whole thing will
be academic. No matter what the contingencies, Ettinger believes, the man who is about to become a corpse will be dead
anyway, so he has nothing to lose by gambling that none of
these contingencies will ever arise.

If a corpse is someday revivable, could there be such a thing
as a "cause of death?" Suicide would be—well, dead. Would
failure to freeze someone be considered murder? Might people
who were bored or unhappy with the times they lived in have
themselves frozen in the hope of coming back to life in a
happier, more congenial era? Might a criminal use the freezer
as a means of escaping punishment? (For that matter, might
freezing become an official means of "execution"—thus satisfy-
ing both those who were for and those who were against
capital punishment?)

Certainly people with diseases now incurable might want to
arrest the damage at the earliest possible moment taking advantage of the new unidirectional time machine to give themselves the best possible chance in the future.

Religion, no less than law, will have to help decide the right
and wrong of all this. Quite apart from the consequent revision
of ethical and moral codes, theological doctrines already in flux
will undergo even severer reexamination. What, for example,
of the soul? With the coming of the Space Age, theologians
have already found themselves constrained to consider whether
or not beings who might be discovered living on other planets
in the universe would possess souls. But the soul on earth is far
from a settled matter. And we need not await the freezer era to
be faced with the same questions. If a man has been dead for
five minutes, for instance, and then brought back to life, where
was the soul in the interim? Did it leave the body and then
return? Or did it remain there the whole time? If so, did souls

also remain in dead bodies in the past? At what point does the soul depart, anyway? In Pirandello's play, *Lazzaro,* the entire population of an Italian village loses its faith because a man who had undergone cardiac resuscitation could not recall having been in paradise in the interim.*

In the freezer age, with so many bodies kept in suspended "animation," the problem will be more pressing, but not actually more difficult. If it can be satisfactorily explained where the soul dwelt for those five minutes, the same answer would serve just as well for five centuries. But what about a body that has been implanted with new organs on a wholesale scale, where even the heart and brain have been replaced? The heart used to be thought of as the seat of the soul. Brain surgeon Robert J. White, who is Catholic, is convinced that the soul—which he thinks of as some mysterious "perfume" or essence of a person's being**—must reside in the brain or head. We have

---

* "An interesting legal problem can likewise be devised," says the eminent British law professor, Glanville Williams, in *The Sanctity of Life and the Criminal Law.* "Suppose that when a rich man's heart stops, and as the physician is about to attempt to revive him, his heir plunges a dagger into his breast in order to make sure that he is not restored to life. Is such an act the murder of a living man, or a mere unlawful interference with a corpse?"

** An Arizona prospector named James E. Kidd recently left a $200,000 estate to be used for research to provide "some scientific proof of the soul." He suggested that a photograph might be made of the soul as it leaves the body at death. More than ninety people and institutions applied for the money, including the American Society for Psychical Research, headed by Dr. Gardner Murphy of the Menninger Institute in Topeka, Kansas. "If the soul exists," said Dr. Murphy, "it is my opinion that it is probably not tangible." However, he goes on, "scientific evidence might show that it is separable from the human body. If it is directly detectable, photographs could be logically expected to give positive results." The judge finally awarded the money to a research institute in Phoenix.

In a slender early novel called *Weigher of Souls,* André Maurois told

already asked if an artificial chimera remains the same person. Does he retain the same soul? If not, where did the old one go, and whence came his new one, if any?

The prospect of tissue-culture reproduction ties the riddle into further knots. If a new being were grown from the cell of a man still living, a possiblity to be elaborated upon later, would it too have a soul? If so, it cannot be the man's own soul because his still presumably remains in residence. Where did the new soul come from? If he has none, is he human? Can he be saved? And suppose one hundred people are grown from the still living cells of a dead man. Do they all have souls? Did the dead man whose tissue provided the cells have one hundred souls to spare? Where were they meanwhile? Perhaps in the DNA of the cell nucleus? If so, do the millions of cells still kept alive in the culture also have souls?

It has been demonstrated that, even in hamburger meat, there are still live cells which, transferred to a congenial medium, will begin multiplying. If a new cow were grown from such a cell, would it mean that the old cow, though butchered and ground up, had never really been killed at all? Applied to people, this means that a corpse mangled beyond recognition in some horrible accident would still yield up intact cells from which a new being might be grown. Or that a frozen cadaver which turned out to be not entirely revivable as a whole human being might still contain cells that would thrive in culture. Dr. Elof Carlson, a U.C.L.A. zoologist, has suggested that even the mummified cells of people long dead might be viable for tissue-culture growth. If any one of these cells has the capacity to become a human being, is there a soul resident in each one?

---

the tale of a scientist who succeeded in capturing this essence or perfume of the soul in a laboratory flask as each of two brothers died. The brothers were very close in life, and the scientist saw to it that their souls were happily commingled in death.

In those theological circles where the death of God is calmly discussed and written about, the disappearance of the soul would scarcely make any waves. But some theologians might sorely miss it. Do our BSP conjurations finally render the soul barren of meaning, and must some other, more up-to-date theological concept be substituted for it? And in that case does individual identity—which has always been thought of as an explicit, unchanging entity—turn out, under the circumstances, to be a fuzzy elusive abstraction?

I have let considerations of the freezer society take me off into flights of fancy far removed from the contemporary medical scene. I now come back to a question so far barely touched on: the really critical human quandaries involved in terminal cases—i.e., patients who are what Dr. Hamlin calls "heart-lung preparations," people who have virtually no chance to recover in any meaningful human sense to any meaningful sort of life.

Dr. DeBakey has experienced this perplexity in those instances when his still experimental artificial bypasses and boosters for the heart have outlasted the patients. A man with his circulation kept going by artificial means can still experience acute failure of the kidneys or liver, or the rupture of a lung—in which case only the continuing circulation is keeping him alive; but alive in what sense? "Should this life-saving device be made available to every patient," asks DeBakey, in the *Journal of Rehabilitation,* "even the hopeless victim of stroke, cancer, or senility? Or should an unbending and restrictive criterion for use be outlined? When and how does one determine death due to other causes? And who decides when to terminate the power flow in such cases?"

The relatively recent ability to keep life going in this manner for short periods of time has given rise to controversy on many levels. The most persistent basic question—as in other current medical situations—has been: Are we obliged to do these things merely because we can? Speaking at a Lasker Award luncheon in New York, Vice President Hubert Humphrey declared, "Every human being deserves every day of life that science can win for him." But a day of what kind of life? Does the mere pumping of the vital juices through an inert and unaware body constitute life? Does it constitute life especially in cases where medical authorities are agreed that there is 100 percent certainty that the patient will die and zero percent chance that the patient can be revived into any meaningful manifestation of sentient life?

Imagine a very old man in this situation. The children and relatives have repeatedly told the doctors, "Spare no expense. Do everything possible." The doctors are in any case committed by their codes of ethics to do everything possible, and they do—except, of course, in those cases where they just quietly decide not to. Imagine that, despite the best that everyone could do, the patient has had repeated strokes and has reached a point which in former days would have been called death. But now a whole array of new techniques have been brought onto the scene: respirators, pumps, heart-lung machines, artificial kidneys, and whatnot. The patient is bristling with tubes and catheters which have been stuck into every natural orifice of his anatomy as well as several manmade ones. He is fed intravenously and hooked up to the appropriate machines while doctors, nurses and expensive medical facilities are tied up, perhaps on a round-the-clock basis, just so a feeble heartbeat can be maintained from minute to minute, from hour to hour, from day to day, with no realistic hope that anything more than that can be accomplished. It is thenceforth a matter of continuing the frantic efforts until finally even those frantic efforts are to no avail, and death by any definition can no longer be postponed.

At this point the doctors feel certain there is no hope, but they do not want to take it upon themselves to stop the treatment. The family may know there is no hope—but who would not feel inhuman and unnatural saying, "Stop, let him die." Meanwhile—*at a point after all hope is gone*—the poor body that is not quite a corpse continues to be mortified, the family's emotional and financial resources are totally drained, and all kinds of medical personnel and equipment that might be more helpfully diverted elsewhere are monopolized by this last-ditch agony.

Speaking at a medical assembly in New Orleans, Dr. Thomas T. Jones of Durham, North Carolina, spoke for many doctors when he voiced his doubts that families want to remember "their loved ones as being riddled with tubes and catheters and tended by machines," and questioned seriously whether doctors should "interfere with the act of dying by unnaturally staving off a death that is coming anyway." Death, he declared, should be allowed to proceed with dignity.

Dr. Frank J. Ayd, Jr., chief of psychiatry at the Franklin Square Hospital in Baltimore, writing in the *Journal of the AMA*, stated his position with regard to both patient and doctor. "An ill person is not obligated to do what is physically or morally impossible. Hence, he can validly refuse treatment which would entail great suffering, which would 'overtax the will power and courage of the normal person,' or which would financially impoverish his survivors, especially if the anticipated benefits would be of limited value and brief duration. Likewise, a physician, with the patient's consent, may licitly desist from administering treatments that are demonstrably ineffective and be satisfied with alleviating the patient's suffering. No one is required to do what is practically useless. A doctor is not under compulsion to make every effort to prolong every patient's life."

Yet, when it was revealed last year that a hospital in London had quietly posted a set of criteria which would guide doctors and nurses as to what kind of patients to save and what kind not to save, such a howl was raised that the guidelines were quickly withdrawn. On a Public Broadcast Laboratory program on Station WNDT-TV, New York, Dr. William Rial of Philadelphia said that there are hospitals where similar guidelines are followed, but it is done quietly and without publicity. A nurse on the same program said that "very often, the decision comes for a nurse to make a choice between the terminally-ill patient and the other patients in the ward. I think very often," she said, "you have to make the choice that the other patients who have a greater chance of survival have to be the ones that are taken care of first. And the terminally ill patients be made comfortable, but left to die." But Dr. David Karnofsky of the Sloan-Kettering Institute insisted: "As physicians, our obligation is to sustain life and to preserve it as long as possible. Secondly, we have a number of young men who train in this institution, and they're taught to do everything possible in the cure of their patients. If we arbitrarily decide that a patient has lived long enough or that nothing further can be done for the patient, we may ultimately make serious mistakes and discourage these men from carrying on in the best possible way."

Later in the same discussion, Dr. Peter Bowman, superintendent of a center for the mentally retarded in Portland, Maine, brought up the similar question of how far to go in trying to keep alive someone who is mentally retarded. He cited some hopeless, untrainable cases, and showed movies of them. First, "a twenty-one-year-old girl, who's severely retarded, she must be restrained at all times. She doesn't feed herself, she refuses to be fed and she has to be force fed . . . Or this fourteen-year-old boy who is a hydrocephalic. He's profoundly retarded. His head circumference is 36½ inches. And he must be restrained. It takes two people to change him . . . This fifteen-year-old boy is totally blind, convulsive, is difficult to feed, extremely self-abusive . . . And here is a twenty-eight-year-old man, operated on several times. He has severe attacks of pneumonia and grand mal epilepsy. Eye tract infection, ulcers. He may live for another forty or fifty years."

"Are you suggesting," he was asked, "that these people be killed?"

"No, I am not," he replied. "I'm raising the question as to what form of treatment these patients should receive when they become seriously ill, whether they should be permitted to peacefully die."

The moral question here is a little different, though, than the decision in regard to a normal person who is altogether beyond any true medical help. "When death is imminent and inevitable," says Dr. Ayd, "it is neither scientific nor humane to use artificial life-sustainers to protract the life of a patient. Instead, when realistic hope of recovery has evaporated, it is the right of the patient to choose only ordinary means to sustain his life, and it is the duty of the doctor to provide palliative care. Only when there is a reasonable hope of sustaining life for several weeks or months, and if during this time the patient can be comfortable, should we exert every effort to delay death. Otherwise life-preserving treatment ceases to be a gift and becomes instead a scientific weapon for the prolongation of agony. As physicians we must recognize the dignity of man and his right to live and die peacefully. If we do not do so, we are failing to strike a balance between the science and the art of medicine, to the detriment of our profession and the degradation of our fellow-men—our patients."

Another spokesman for letting people die with dignity is Dr. Irvine Page. "A man," he says, in *Modern Medicine,* "should still have the right to die at home among those he loves. All reasonable efforts should be made to stay death, but there are unreasonable limits as well." Page wonders if physicians really do have the responsibility to "keep the body alive after it has irretrievably failed." "Do we," he asks on another occasion, "want to keep people alive beyond their span, to deny their rights to die, at a cost of resources and human endeavor taken from something else?" And again: "How much of our substance can be given to prolonging artificially the lives of a few, while denying life to the many who still have it to live?" Dr. Page of course understands that it is easier to enunciate such principles than it is to make the critical decision in a specific case. "When do you stop? Do you in fact decide on life or death? Is this homicide?"

Discussing this problem at a Yale University medical symposium, a well-known rabbi, Dr. Isaac Klein of Buffalo, quoted a Jewish authority to the effect that "if the physician refrains from healing, he is guilty of shedding blood." This would seem to equate the decision Dr. Page is talking about—the decision to turn off the machine—with euthanasia. Doctors do, as has already been pointed out, make this sort of decision all the time. But they do it at their peril. In a historic euthanasia case in Sweden, Dr. Olle Salin was tried because he had discontinued, with the children's consent, the intravenous feeding of an eighty-one-year-old victim of a cerebral hemorrhage who, had the doctor not made this decision, would probably have lasted a few more days. Dr. Salin was finally found innocent, and Swedish newspapers applauded the decision. But other courts in other places might decide differently.

At the same Yale symposium, the Reverend Joseph Fletcher of the Episcopal Theological School, Cambridge, Massachusetts, and author of *Situation Ethics,* made a careful distinction between euthanasia and what he calls "anti-dysthanasia:" "This latter form is different from the classical concept of euthanasia because it is a procedure whereby, even though death is brought about quite rationally and deliberately, it is accomplished only indirectly through omission rather than directly by commission. It is, in short, a procedure by which death is not induced but only permitted." He concludes that it is perfectly consistent with Christian ethics to believe that "there is no absolute obligation to preserve a patient's life simply because it is medically manageable to do so." A surprising number of theologians are in accord.

As Bishop Fulton J. Sheen put the question at a meeting of the AMA's Committee on Medicine and Religion, "Should a patient's life be prolonged to exist only as a kind of vegetable, with no hope of ever becoming a total life?" And he suggested that, under some circumstances, a valid answer might be No. Dr. Paul S. Rhoads, a Chicago physician, said in that same discussion, "The obligation of the physician is to do the best he can for the patient. And sometimes the best thing he can do is let the patient die."

At another AMA meeting, Dr. Williamson commented that "intensive efforts to maintain life can become prolongation of dying rather than prolongation of living." The dilemma is made a bit stickier, suggested the Reverend Granger E. Westberg of Houston, if the question is whether to let one person die in order to save another. Suppose you have a terminally ill patient being kept alive by machine—the only one you have— and another patient turns up. This one has a good chance to get better—but, to do so, he needs the machine. Do you transfer the machine? And, if you do, have you in fact killed the patient by taking it away? "Everybody is willing to turn the machine on," said Dr. Westberg, "but nobody is willing to turn the machine off. The doctor needs some help in making this decision."

Dr. Williamson, writing later in the *Journal of the AMA,* agreed: "It is proper for us to be merciful, but we must be human about it, and not pretend to be omniscient. It is always unwise for the doctor alone to make the decisions. Unilateral edicts are dangerous. Certainly, the patient and the family have some say in these matters as well as others meaningfully involved in the particular situation. Basically, it often becomes a religious decision."

Though the matter is by no means settled, the consensus seems to go along with poet Arthur Hugh Clough who, in *The Latest Decalogue,* advised: "thou shalt not kill; but need'st not strive/ Officiously to keep alive."

All too often, it seems likely that extraordinary life-preserving measures—experimental surgery, for instance—are undertaken more to advance knowledge in the ultimate hope of saving *future* patients than with any realistic hope of saving the patient of the moment. But "a doctor must recognize," says Dr. Ayd, "that he does not have the right to urge a dangerous remedy or a new procedure without just cause. However desirous he may be of learning, of making progress, or of doing something for the common good, the practitioner must not yield to the temptation to 'sell' his proposal in order to acquire the necessary authorization. All men have inalienable rights which transcend any considerations of science or of the good of others."

Doctors have been severely criticized by their colleagues for implanting in human patients synthetic parts, monkey kidneys, or chimpanzee hearts before there was any certainty that the implants would work. An editorial in *The Annals of Internal Medicine,* the official journal of the American College of Physicians, took to task surgeons who had performed lung transplants in three patients when, in the opinion of the *Annals* editors, none of these patients had the remotest chance of surviving the operation. In undertaking any transplant operation, said Dr. Eugene D. Robin of the University of Pittsburgh School of Medicine in the *Journal of the AMA,* "the transplant should have some reasonable possibility of clinical success." This led him to the conclusion that heart transplants should not be done at all. "It has been well established experimentally in the animal that cardiac homotransplant cannot maintain life on a long-term basis. Therefore, he reasoned, "there is no moral justification for attempting this procedure in human patients. Nor should the physician rationalize such attempts by the consideration that the patient 'was going to die anyhow.' The fundamental role of the doctor must be that of healer and not of executioner."

Those words were written some time before Dr. Barnard performed his historic operation. In the interim, encouraging progress had been made in immunosuppressive techniques in connection with kidney-transplant patients—enough to enable Dr. Robin to agree with others, like Drs. Moore and Murray, that a cautious beginning could now be made in heart transplantation. But he also agreed with Dr. Page that only a limited number of such operations should be done and then a moratorium called until the results of the first half dozen or so could be evaluated. In accord, too, was Dr. Ramon Suarez, a cardiologist of San Juan, Puerto Rico, who warned against "an epidemic of heart transplants." The "epidemic" occurred anyway. Heart transplants were done from Brazil to India—with Dr. Denton Cooley performing three in one five-day period at the St. Luke's Episcopal Hospital in Houston.

On the other hand, Dr. Reed—as cautious as anyone—wondered whether a decision on the side of over-caution did not carry on equal burden of ethical responsibility. "Suppose the right combination of circumstances comes to pass," he said, "with recipient and donor both answering the most rigid criteria, and both of them on hand at the right moment, in an institution which has a team of experienced open-heart surgeons as well as a competent backup team equipped with the latest and best immunological knowhow. This combination would occur only rarely, but when it does, and this is the patient's only chance, must we not ask ourselves: Do we have a moral right *not* to try a transplant?"

On the whole, though, medical opinion seemed more critical than not. Dozens of prominent doctors in the United States and abroad were willing to denounce the heart operations as premature in the present state of knowledge, and sentiment in the Soviet Union was so strong that Health Minister Boris Petrovsky explicitly forbade their performance. Most extreme, again, was Dr. Forssman, who pronounced it "a crime to perform an operation in a field where fundamental research is not yet finished."*

It is not that these doctors were worried about the state of surgical knowhow—though there were some fears expressed that surgeons with insufficient experience or facilities might be tempted to try it. And there was undoubtedly public confusion on this score, because, to the layman, initially it must have seemed that the surgery performed by Dr. Barnard was a truly miraculous feat. The aura emanating from the various surgical theaters at the time did somehow turn out, inadvertently, to be more evocative of showbiz than of science—so much so that Dr. Kantrowitz was moved to deplore the "circus atmosphere" in which the operations had been carried out.

It was perhaps inevitable that the first heart transplant in history would create a sensation. The sheer, breathtaking daring of it made for human drama on an epic scale. An irresistible story. A medical miracle. A major publication at the time (under the headline MIRACLE IN CAPETOWN) called the feat

* A scant year after Dr. Barnard's first operation in Capetown, nearly 100 heart transplantations had been performed around the world despite continuing vigorous criticism. In two of these cases, at least three other organs were removed from the heart donor. Toward the end of 1968, I attended two medical meetings—one in New York, one in Miami—where transplant surgeons freely expressed their readiness to remove still-beating hearts from donors once "brain death" had been ascertained. One heart transplant was even finally done in the U.S.S.R. where the operation had at first been banned.

"perhaps the most remarkable bit of surgical virtuosity in the history of medicine." But it was something short of a miracle. The surgical performance was unarguably splendid, but not uniquely so; and the operation, though unquestionably difficult, was not *the* most difficult in history, nor the most difficult of the century, nor even, probably, the most difficult done that same day.

As Dr. Reed explained it, the transplant is "a relatively simple operation" for a team of experienced open-heart surgeons. Any automobile mechanic knows that it is sometimes easier to replace an entire motor than to do a lot of intricate repair work within it. Dr. Robert J. White has explained how much easier it would be to transplant an entire head than to take out a brain and transfer it to another head. In the same manner, it is easier to replace an entire heart than to do extensive repairs inside it—the kind of repairs open-heart surgeons are called upon to do routinely.

Hundreds of heart transplants had been done experimentally in dogs, and techniques carefully worked out for duplicating the feat in human patients. Dr. Shumway had in fact predicted that it would require less suturing skill to do the operation in people than in dogs because of the greater fragility of the dog's large blood vessels. In short order Dr. Cooley had worked out a simplified technique that enabled him to sew a new heart into a chest in thirty minutes. So there was no question of prematurity with regard to the surgery. The objections were all based on the fact that the rejection problem still remained unsolved. Despite the rather hopeful advances made in kidney transplants (with some 40 percent of the recipients of cadaver kidneys now surviving for a year or more), it is impossible to extrapolate methods directly from kidney patients to heart patients. They are quite different organs with quite different problems. Some laboratory evidence, for instance, indicates that hearts may be rejected less readily than kidneys; other data suggest that heart-transplant patients may, for reasons still obscure, be less able to tolerate the drug dosages commonly administered in kidney cases. There is only one way to find out with any certainty what needs to be known: experience with human patients. And the first patients provided the necessary experimental subjects.

Every new operation is experimental until—well, until it no longer is experimental, and there is no clean-cut, accepted set of criteria for establishing precisely when that point has been reached. There was a celebrated decision handed down in New York in 1871 (*Carpenter v. Blake*) in which the court ruled that the law "admits the adoption of new remedies and modes of treatment only when their benefits have been demonstrated." But how demonstrate the benefits before trying the therapy? No amount of laboratory or animal research offers any assurance that the operation will work on people. If it works on one person, there is no assurance that it will work as well on the next. People still do die, after all, of operations which everyone considers to be routine minor surgery. The distinction between experimental therapy and what every doctor does every day is fuzzy at best. In a sense, every patient is every doctor's experimental subject, since the doctor can never be absolutely certain that his diagnosis is correct, that his treatment will be beneficial, or even that it will not be harmful.

When may a surgeon feel free, then, to attempt a so-called experimental operation on a real human patient? As a rule, he may do so only after every other treatment has failed—when, without this last-ditch try, the person would surely die. But by that time, of course, the patient is so far gone that he is a poor all-around risk, and the new technique—whatever it may be—does not really get a fair test. This last handicap appears to be unavoidable. Some doctors feel that such patients should simply be allowed to die in peace. But it is generally accepted that, as long as the patient and/or his next of kin have been given a

85

realistic view of his chances, and the hardships he will undergo, the surgeon is justified in proceeding.

Experimentation with human subjects is a continuing subject of medical concern. The debates range from the testing of drugs to the implanting of electrodes to the most complex diagnostic and therapeutic techniques. Among the more vocal critics have been Dr. Beecher in the United States and D. M. H. Pappworth, author of *Human Guinea Pigs,* in Great Britain. As more and more tampering becomes possible with the body's most basic processes,* these questions become more insistent than ever. Under what circumstances and under what kinds of controls should any patient be subjected to an unproven treatment? When does an experimenter feel safe in inseminating a woman with frozen sperm, in administering a drug with toxic side effects, in performing a liver transplant?

And how carefully and thoroughly must the experiment be explained to the subject before the requirement for "informed consent" can be said to have been met? In 1966 two highly respected New York physicians were found guilty of "unprofessional conduct" and of "fraud and deceit in the practice of medicine" because they had failed to make it clear to a group of elderly hospital patients that the injections they were receiving contained a suspension of cancer cells. Though the research had produced useful results, and at no real risk to the subjects, this did not let the researchers off the hook, in the opinion of the Regents of the University of the State of New York, who have the responsibility for licensing doctors in the state. In their decision, the Regents ruled:

---

* This is not by any means restricted to the academic research laboratory. In the biological labs of frankly profit-oriented Dupont, the stated aim is "to interfere with basic biological processes in beneficial ways." *Interfere* is the key word. Another word is *beneficial,* but once you learn to interfere, you can do it either way.

No consent is valid unless it is made by a person with legal and moral capacity to make it, and is based on a disclosure of all material facts. Any fact which might influence the giving or withholding of consent is material. A patient has the right to know he is being asked to volunteer, and to refuse to participate in an experiment for any reason, intelligent or otherwise, well-informed or prejudiced. A physician has no right to withhold from a prospective volunteer any fact which he knows may influence the decision. It is the volunteer's decision to make, and the physician may not take it away from him by the manner in which he asks the question or fails to explain the circumstances. There is evidenced in the record of this proceeding an attitude on the part of some physicians that they can go ahead and do anything which they conclude is good for the patient, or which is of benefit experimentally or educationally and is not harmful to the patient, and that the patient's consent is an empty formality. With this we cannot agree.

It is not that anyone really disagrees with the basic principle that the physicians must not withhold any information the subject might need to arrive at a rational decision, or omit any mention of the risks. But there are various ways to interpret this principle. Some people feel it is possible to go to the other extreme and tell a patient too much. Dr. David W. Louisell, professor of law at the University of California at Berkeley, believes that "to explain every risk, no matter how remote, could produce bad medical practice by requiring too much talk, thus alarming a patient perhaps already unduly apprehensive."

The possibility of overexplanation was illustrated in a lecture given at Yale by Dr. Jay Katz, a distinguished physician and law professor there, who told of a doctor's imaginary conversation with a patient, Mrs. Blair, who needed a hysterectomy. The doctor, in line of duty, was recounting all the things that could possibly go wrong. "He told her about the possibility of nurses nicking her skin while she was being shaved, or cystitis

87

on catheterization, of drug reactions, of viral infections and hepatitis resulting from venipuncture, of intestinal perforations and subsequent peritonitis from enemas, of anesthetic cardiac arrest with resulting brain damage, of intestinal and ureteral perforations from surgery, of postoperative breakdown of sutures, etc. . . . . 'Mrs. Blair! We're not through. Where are you going? . . . Mrs. Blair!' "

Hardly anyone opposes medical research or resents the money spent on it, for we all hope to reap the benefits. Yet, if medical progress is to be made, pure research is not enough. A theory may be proven in the laboratory. A potential therapy may be tested *in vitro,* in tissue culture, in animals. But there reaches a point where someone must dare to try the new treatment on human beings.

Some people talk and act as if they believed medical progress was possible without human experimentation. But how? Nothing could be more obvious than the fact that, if you can't use a new therapy on people, then people can never use the therapy. And there is no such thing as a riskless human experiment. No matter how painstaking the lab work, the experimenter can never be absolutely sure, when he gives a man a pill never before swallowed by any human being, that the subject will not drop dead on the spot. In general, the public is aware that risks must precede the benefits.

Dr. Katz puts the problem in a larger social context.

> Society tolerates constant experimentation with, for example, insecticides, food additives, atomic explosions, psychological tests on school children, without being clearly aware of its implications and therefore, medical experimentation is indeed only another example of our experimental way of life. Also society is willing to underwrite the human cost of such experimentation. Lon Fuller, the distinguished Harvard professor of jurisprudence, stated the issue well: "Every highway, every tunnel, every building we project involves a risk to human life. Taking these projects in the aggregate we can

calculate with some precision how many deaths the construction of them will require; statisticians can tell you the average cost in human lives of a thousand miles of a four-lane concrete highway. Yet we deliberately and knowingly incur and pay this cost on the assumption that the values obtained for those who survive outweigh the loss."

It is one thing, however, for society to take statistical risks—to decide on a project or a program even though the probability is high that a certain number of people will suffer inconvenience, discomfort, illness, injury, or death. In such cases, no one can predict who, precisely, the unfortunates will be; at the outset, they are represented only by abstract statistical probabilities. But medical experimentation is different. Here the experimenter himself *selects*—or asks to volunteer—the specific individual human beings he will submit to the risks. Thus a much more personal kind of decision and responsibility are involved. Does it require a special kind of mentality to undertake this kind of decision and responsibility? Dr. Katz quotes a famous physician who said, "The desire to alleviate suffering is of small value in research—such a person should be advised to work for a charity. Research wants egotists, damned egotists, who seek their own pleasure and satisfaction, but find it in solving the puzzles of nature." Katz warns that "the researcher's interaction with a patient becomes significantly determined by his quest for knowledge, and the care of the patient becomes a secondary, or at least a conflicting consideration."

Is it ever, under any circumstances, fair to ask some people to undergo unnecessary hazards to insure the later well being of many other people they will never get to know? The late Dr. J. B. S. Haldane always thought so. "I have often risked other peoples' lives in physiological experiments," he said at a Ciba symposium in London in 1962, "and though none died, at least one was permanently injured. But they were all volunteers, and I was taking the same risks as they. The exploration of the

89

interior of the brain will be as dangerous as that of the Antarctic continent or the depths of the oceans, and far more rewarding. The 'officer in command' must be a man of proved personal courage, but not so soft hearted as to leave his post of command because his orders have led to some deaths, mutilations, and psychoses."

A scene from Sidney Howard's *Yellow Jack* illustrates what is often the researcher's attitude—fascination with phenomena rather than people:

GORGAS (*low and quick*): "A hundred and three and six tenths last night. Dropped again, though, at six this morning and again at eight."

FINLAY: "You've noticed the granular casts in the urine, I hope?" (REED *goes toward them.*)

GORGAS: "Oh, yes."

FINLAY: "The eyes were beautifully jaundiced today, too."

GORGAS: (*to* MISS BLAKE) "How about the gums?"

MISS BLAKE: "A little bleeding."

FINLAY: "Headache and nausea still troublesome, though?"

MISS BLAKE: "He's very uncomfortable."

FINLAY: "Splendid! I should defer to the Major's diagnosis, but I can't think of a symptom the boy's omitted! It's beautiful! Beautiful! The fourth day of his sickness, too!" (*Then, to* REED.) "And how long did you say between the bite and the first symptom?"

REED: "Three days, nine and a half hours."

FINLAY: . . . (*He is pumping* REED's *hand.*) "I conceived a truth! You delivered it into life! Together we have added to the world's arsenal of knowledge!"

90

The Refabrication of the Individual

Or a passage from Sinclair Lewis's *Arrowsmith:*

> It comes to me (said Gottlieb) that there is pneumonic
> plague in Manchuria and bubonic plague in St. Hubert, in
> the West Indies. If I could trust you, Martin, to use the
> phage with only half your patients and keep the others as
> controls, under normal hygienic conditions but without the
> phage, then you could make an absolute determination of
> its value, as complete as what we have of mosquito trans-
> mission of yellow fever, and then I would send you down to
> St. Hubert. What do you t'ink?
>
> Martin swore by Jacques Loeb that he would observe test
> conditions; he would determine forever the value of the
> phage by the contrast between patients treated and untreated,
> and so, perhaps, end all plague forever; he would harden his
> heart and keep clear his eyes.

Few researchers would state this philosophy as bluntly as
Haldane did. It is a philosophy which, when stated thus
bluntly, scares people. It makes them wonder if there is not
after all some truth in the characterization of the mad scientists
of fiction, TV, and comic strips—the scientists who, pursuing
their own abstract truths, look upon people as mere experimental objects to be manipulated, measured, and observed with
cold objectivity. The dehumanization that can result from this
viewpoint, carried to its ultimate, is nowhere better exemplified
than in the Nazi experiments of World War II. Here is a
sample description, and certainly not the most horrendous
available, of one such experiment wherein the cool objectivity
of the experimenter is not to be doubted:

> The third experiment . . . took such an extraordinary
> course that I called an SS physician of the camp to witness,
> since I had worked on these experiments all by myself. It
> was a continuous experiment without oxygen at a height of
> 12 kilometers conducted on a 37-year-old in good general
> condition. Breathing continued up to 30 minutes. After 4
> minutes the experimental subject began to perspire and

91

wiggle his head, after 5 minutes cramps occurred, between 6 and 10 minutes breathing increased in speed and the experimental subject became unconscious; from 11 to 30 minutes breathing slowed down to three breaths per minute, finally stopping altogether. . . .

Severest cyanosis developed in between and foam appeared at the mouth.

At 5 minute intervals electrocardiograms from three leads were written. After breathing had stopped, EKG was continuously written until the action of the heart had come to a complete standstill. About ½ hour after breathing had stopped, dissection was started.

If you adopt a philosophy that knowledge is to be pursued no matter what, then this is where you wind up. You want to know the exact sequence of physiological events when a man dies under a certain set of conditions? Then simulate those conditions—put an experimental subject (never call him a "man") in the required context, and let his breathing stop (never using the word "death"), meticulously observing and recording his reactions in great detail. Thus putting to death becomes science rather than murder.

The Nazi philosophy is admittedly extreme, and is almost universally condemned as being sickeningly, psychotically inhumane. Yet even the Nazis could only rationalize their experiments by imputing an essential racial inferiority to their subjects. Once the subhumanity of a category of people is accepted, the next step is not too hard to take. Most of us accept the premise that it is all right to inflict horrible diseases upon thousands of healthy animals—even breeding them for that specific purpose—in the hope that the experimental findings will save some human lives. That inferior beings may be sacrificed to save superior ones, then, is a logical extension of such a rationale. In the past it was not a rarity for doctors to apply this idea to people. If an orphan, or the child of a peasant or a prisoner, had to be given smallpox to provide material for the vaccination of a prince, what royal physician would hesitate?

Any such practices are certainly deplored in the contemporary world, as they are bound to be in any milieu that preaches democracy and egalitarianism. But have we altogether graduated from such notions? In looking for experimental populations, is it not a little easier, more acceptable, for scientists of a technologically advanced nation to try out a new therapy somewhere in a "backward" nation (with that nation's approval, of course)? Or to seek a group of volunteers in some institution—a prison, an orphanage, a home for the aged? Charity hospitals have been a favorite site for the proving-out of experimental techniques, often painful and dangerous, and often having little or nothing to do with the disease of the subject being experimented upon. Some of the instances cited by Dr. Pappworth in his book are incredible enough to invite comparison with what the Nazis did. Often sick persons have deliberately been made sicker in order to study the illnesses. And Pappworth is openly skeptical of the means of obtaining consent. He feels reasonably certain, too, that the most egregious cases go unreported; his case rests almost entirely on reports by the doctors themselves in the technical literature.

Dr. Katz observes that "we have been satisfied with fulfilling legal standards rather than asking ourselves whether they conform to our own ethical standards. For example, we have experimented* on mentally retarded children after scrupulously obtaining consent from the administrators of the institutions without deliberating sufficiently whether or not it was ethical for us to proceed, especially when the experiments were unrelated to diseases for which these children were hospitalized. Indeed, we may have become so preoccupied with what is

---

* Dr. Katz's "we" here does not apply to any experiments in which he himself participated. It is rather an editorial "we" applying to experimental medicine in general.

legal that we have neglected to define our position from our own vantage point."

A bizarre proposal for circumventing many of these dilemmas and for speedily acquiring a great quantity of knowledge about basic human physiology, has been put forth by Dr. Jack Kevorkian of the Pontiac General Hospital in Pontiac, Michigan. Kevorkian would like the legislature of some state that still permits capital punishment to offer the condemned man the choice of being executed in the usual prescribed manner—or of being placed under anesthesia, never again to awaken, while a skilled medical team used his body and brain for experimentation and study. By using condemned criminals in this manner—men whom society has in any case ordered put to death, and who would thus be offered an opportunity to expiate their crime by making a major contribution to human knowledge—Kevorkian believes that more could be learned in a single year in a single state than is now gleaned in decades of worldwide efforts. Aware of how much his proposal, when looked at superficially, smacks of the ghastly Nazi experiments, Kevorkian has written a book—printed at his own expense—spelling out the differences. He has talked to condemned prisoners, prison wardens, criminologists, medical researchers, and state legislators about his plan, believes he has found a lot of sympathy for it, and still hopes he can convince someone to carry it through.

There are strong dissenters from this point of view. Dr. Page, for one, has grave moral misgivings about what he considers the cavalier attitude we have lately taken toward the human body, which he still looks upon as the sacred vessel of man's soul and spirit. We tend to meddle with it too lightly, he feels, and a disregard for the human body can easily lead to a similar disregard for human life in general. He believes that, rather than push precipitately into experiments that entail unnecessary risks, research should go more slowly—and that the application of its results to human beings should go more slowly still.

In the nineteenth century, the great physiologist Dr. Claude Bernard had to spell out, in a work that is still a classic, a rationale to justify practicing experimental medicine at all. Now we in the latter part of the twentieth century must start from scratch to rejustify experimental medicine on new ethical and moral levels, in consideration of the radical consequences that inhere in the new biology. In 1967, after a long period of study by a group of distinguished academicians and administrators under the chairmanship of the Right Reverend Henry Knox Sherrill, the Board of Trustees of the Massachusetts General Hospital published a booklet called *Some Guiding Principles for Human Studies*—which is expected to provide exactly that for its staff. One of the principles enunciated: "An even greater safeguard for the subject than consent is the presence of an informed, able, conscientious, compassionate, responsible investigator, for it is recognized that subjects can, when imperfectly informed, be induced to agree, unwisely, to many things. It may be trite to say it, but firm application of the Golden Rule would go very far to eliminate difficulties here." Though admitting that "it is not possible to spell out all contingencies," the MGH group did spell out procedures and considerations which other institutions might want to follow. On a broader scale, in 1966 the AMA and seven other American medical societies officially endorsed the ethical principles concerning human experimentation adopted by the World Medical Association in Helsinki in 1964.* They took this

---

* The Declaration of Helsinki consisted of three major sections. The first was a set of basic principles, such as: "Every clinical research project should be preceded by a careful assessment of inherent risks in comparison to foreseeable benefits to the subject or to others." The second part dealt with experiments on patients—e.g., "The doctor can combine clinical research with professional care, the objective being the

action, fortunately, with no feeling of finality, but rather recognizing that the Declaration of Helsinki (like the Nuremberg Code that preceded it) is simply a good start—a jumping-off place for the continuing critical examination and debate, the refinement of principles and philosophy that must evolve as medical science itself goes forward. Even at best, of course, there is no guarantee that forward is the direction in which medical science will go. It is possible to guarantee the opposite, however, for if we do not take thought, the direction we go, including the moral direction, will surely be backward.

If the carrying out of medical research produces problems, the application of the fruits of that research multiplies them. When brilliant new therapies are devised, how are they to be made available to every patient, or even to every medical center, that needs them?* Medical care everywhere is handicapped by a host of shortages and inadequacies in hospital facilities, equipment and trained personnel. It does not do much good, for example, to devise sophisticated blood tests to diagnose a disease if the people who administer the tests lack the requisite skills and sophistication to carry them out properly and the result is misdiagnosis. It is not unheard of for a man to die

---

acquisition of new medical knowledge, only to the extent that clinical research is justified by its therapeutic value for the patient." And the final part spelled out circumstances for experimenting with non-patient volunteers, emphasizing the "duty of the doctor to remain the protector of the life and health of that person on whom clinical research is being carried out."

* When an artificial kidney—i.e., a dialysing machine—was developed at the University of Washington and installed at the Swedish Hospital in Seattle, it was immediately obvious that only a handful of the potential victims could be handled. To help select the few who were to be saved out of the many who were certain to die, the King County Medical Society appointed a committee of seven laymen.

unattended, of a hemorrhage following a tonsillectomy—while, at the same moment, in another wing of the same hospital, quantities of surgeons, nurses and technicians, bolstered by quantities of apparatus, may be engaged in saving the life of another man in a marathon of open-heart surgery. A hospital may be acquiring the latest in cobalt bombs and hyperbaric chambers while unable to keep the bedpans emptied; and its rooms may cost more than luxury hotel suites at the same time its employees are paid the barest subsistence wages. Dr. John Knowles of the Massachusetts General Hospital has been among the more vocal of those hospital administrators who warn that medical care will bog down in total frustration, no matter how spectacular the research achievements, unless the social and economic considerations get equivalent attention. His prediction, though, is optimistic: "I am firmly convinced that the last half of this century will be notable for its attention to the social problems of medicine and that triumphant solutions will equal biologic discoveries in excitement and importance."

What will be the nature of some of these "triumphant solutions"?

A team of doctors at Georgetown University Medical Center has completed plans for a revolutionary new kind of hospital based largely on the key insight that, at any given time, only 20 percent of the patients are really sick, while the other 80 percent are merely undergoing diagnostic tests or are well on their way to recovery. The hospital attempts to provide adequate medical attention for 100 percent of the patients, including the 80 percent who don't need it, as well as adequate hotel and restaurant facilities for 100 percent of the patients, including the 20 percent who are too sick to care. So the already existing Georgetown Hospital will continue to provide relaxed and comfortable facilities for the 80 percent, while a new, streamlined, computerized Georgetown Hospital will provide total, intensive care *at all times* for *every one* of the 20 percent who really need attention.

As 1966 drew to a close, Dr. John Parks, Dean of the George Washington University School of Medicine delivered a paper to the American Association for the Advancement of Science on "Trends Toward Medical Practice in 1975," while Dr. Oscar Creech of Tulane wrote an article in *Medical World News* on "The Shape of Medicine in 1990." Both emphasized the increasing complexity of medical practice with technological advance, and the consequent trend toward the use of computers to simplify some of the complexity, and toward teams of doctors because no single man can know enough to practice medicine all by himself.

Making the bold leap to 1990, Dr. Creech predicts:

> Almost all medical practice will be carried on within medical schools and schools for other health personnel, general and specialty hospitals with facilities for the graded care of patients, outpatient clinics for the diagnosis and care of ambulatory patients, and convalescent and rehabilitation institutes. The private practice of medicine will no longer exist as we know it. Physicians will be full-time employees of the medical center complex, within which they will provide total medical care for the residents of the community. . . .
>
> All diagnostic procedures and some parts of the physical examination . . . will be performed automatically and interpreted by computer systems. The patient's entire medical record from birth will be instantly available. . . .
>
> Medicine will be practiced on an assembly-line basis, but there is no reason to believe that patients will complain much about the loss of the legendary doctor-patient relation.* After all, no one has complained much about the

---

* On the contrary, when *Life* ran an article describing some of the computerized features of the new Georgetown Hospital, many readers wrote in to express their fears about losing touch with human therapists and being placed at the mercy of computing machines, no matter how efficient. People might of course, in time, grow accustomed to the idea.

disappearance of the personal attention we used to get from the grocer, baker, cleaner, and others.

Dr. Creech, surprisingly, disagrees with current cries about the urgency of training ever more numerous armies of doctors to meet the shortages that already exist as well as the greater ones expected:

> What will be needed is a new category of medical personnel which, for want of a more exact title, I shall call clinical associates. After high school these persons would have four years of lectures, demonstrations, and on-the-job training. An additional year of apprenticeship or internship and they could assume the functions performed by most practicing physicians today.
>
> Today's superspecialists will become systems specialists. For example, cardiologists and cardiovascular surgeons will become specialists in cardiovascular disease and will treat all disorders of the heart and blood vessels, regardless of whether they require digitalis or replacement of the mitral valve. Similarly, urology will merge with nephrology, and gastroenterology with gastrointestinal surgery. Medical manpower needs will be something like one system specialist for each 10,000 persons, one general medical specialist (today's internist) for every 5,000 persons, and one clinical associate for each 500 persons.

Dr. Creech's ideas are meant as suggestions rather than solutions, direction-pointers to indicate the kind of leaps that must be made over present categories of thinking in order to cope with the new categories of need. Doctors like Knowles, Parks, and Creech are urging their colleagues to push their imaginations into these new categories.

To keep abreast of new developments, to predict their consequences, to make value judgments, to set national standards, to help direct research into proper channels and see that its fruits are properly applied, Dr. Page has been editorializing and lobbying for the establishment of a National Academy of Medicine that would do for American medicine what the National Academy of Sciences does for American science.

It is fitting that doctors and scientists, who are most closely involved in the new developments, are the first to express their concern. But everyone will have to be concerned. It would be hard to exaggerate either the challenges or the opportunities for educators, for business leaders, for legislators and jurists, for artists and writers, for theologians and philosophers—and for you and me personally. We cannot expect others to do all the worrying for us, or make all the decisions for us. The time is wild and uncharted. No one has been there, so there are no experts. Each of us, whose body and brain may be modified or whose descendants' characteristics may be predetermined, has a vast personal stake in the outcome. We can guarantee that good will be done only by looking to it ourselves.

PART II

The Exploration of Prenativity

THE NONDESCRIPT OBJECT UNDER GLASS MIGHT HAVE BEEN A BLOB of blubber. Yet it commanded the rapt attention of the man in the close-cropped beard. The blob was in fact a tiny embryo. Unlike the pickled-and-preserved embryos on display in the halls of museums and medical schools, this one pulsated with life—with human life. At least, its origins were human. It now lived and grew inside an artificial "glass womb" which the bearded gentleman himself had devised. More than that, he had been responsible for the conception of the child—if it could be called a child. In a sense, he had fathered it. It might even be said that he had mothered it.

According to his account, he had begun with a female ovum removed surgically at just the right, ripe moment. After immersing the ovum in amniotic fluid from a real womb he had then admitted male sperm, and one of the teeming millions of spermatazoa had proceeded to fertilize the egg, just as it might have done in the fallopian tube of a real mother. Under his careful laboratory maternity, the egg had grown into an embryo. The embryo was still alive and growing there in its transparent manmade uterus, the cells proliferating even as he watched.

Until the past decade or so any scene like this would have been out of a bad science-fiction novel. But this is reality—the place, No. 3 Via de Ruini, Bologna, Italy. The time, 1959. The man, an Italian experimental scientist and surgeon, Daniele Petrucci.

The first press reports of Dr. Petrucci's work created a furor in Italy. Vatican sources as well as laymen roundly denounced Petrucci. An outraged citizen of Naples even demanded that the doctor be prosecuted for murder because he "terminated the experiment"—i.e., let the embryo die—at the end of twenty-nine days.

Dr. Petrucci has reported that one of his *in vitro* embryos survived for fifty-nine days.

103

Petrucci, a Catholic, was naturally upset by the outcry. He has given up growing embryos and has gone on to less controversial experiments, claiming that he never intended to raise a baby in this fashion but had merely been interested in learning how to grow embryonic tissue for possible use in transplantation. Some believe Petrucci may still be carrying out his work in secret. Others regard him as a fraud, insisting, despite the impeccability of his medical credentials—and despite his presentations, complete with movies of his purported embryos, at international medical meetings—that he has never really grown any embryos *in vitro* at all, and that the scene, as conjured up, might just as well have been science fiction after all. Whether Petrucci did perform as advertised, the relevant point is that what Petrucci says he achieved is perfectly feasible. Other researchers, including at least two Americans—Dr. John Rock at Harvard and Dr. Landrum B. Shettles at Columbia Presbyterian Hospital in New York—had already grown embryos *in vitro* before him, though their embryos lasted for only a few days each. Their results, too, were met with some skepticism, but Dr. Cecil B. Jacobson—himself a skeptic of earlier claims—now seems to be indisputably growing early embryos *in vitro* in his laboratory at George Washington University. Many scientists, including a number of Russians, have continued to probe this area of research, and they make it plain—though the details they supply are skimpy—that they have more in mind.

It was not until well into the seventeenth century that anyone caught sight of that implausibility, a human spermatozoon, under a microscope. And another two centuries went by before scientists began to acquire a dim understanding of just what happens inside the female when a human mating takes place—how the sperm penetrates the ovum, and the role played by each in the reproductive process.

The architecture of our morality and the structure of our

tradition rest solidly upon the foundation of reproductive biology. The study and practice of obstetrics and gynecology, begun in earnest not much more than a half century ago, constitute one of the younger branches of medical science. But the generations that lived on earth before the onset of these investigations and discoveries, though they were ignorant of the precise mechanisms, understood the basic and hitherto unarguable facts: that a man and a woman must unite sexually in order to produce a child; that the child somehow begins to develop, on its own and in secret—inaccessible except for whatever mysterious interchanges take place between itself and its mother—for the long, dark, quiet months before it is ejected from its snug place into the shock of life outside; that the helpless mammal that is a human infant requires an unusually prolonged period of parental protection and training before it can cope on even a minimal basis with either its physical or its cultural environment. An infant horned toad bursts forth from the maternal sac all ready to fend for itself. A newborn giraffe or zebra can run beside its mama within a very few hours. But the human baby is helpless.

All this being so, and assuming that human beings in even a relatively primitive state would instinctively want to insure the perpetuation of their kind, it was inevitable that certain sets of conventions would evolve. Thus grew our social institutions of marriage and the family, buttressed by religion, law, politics, philosophy, education, commerce, literature, and the arts—an interdependent edifice of formidable dimensions, endowed by its creators with an aura of self-evident immutability. For the rearing of the young, there had to be some continuity of place, and the assignment of responsibility. The mother could not give her baby the tender care and attention it needed to survive, and also be the one to fight and protect, to feed and clothe and shelter. That meant the father had to be discouraged from straying. Society channeled sexual urges toward one goal—pro-

105

creation—devising complicated systems of prescriptions and proscriptions. Everyone in the family, from infants to uncles to grandmothers, had his assigned role, knew what was expected of him, was aware of his rights and duties and privileges. Courtship was ritualized, wedding vows solemnized, family support enforced. Parents were to be obeyed, elders to be respected, children protected, spouses to be the exclusive sexual property of one another. And theologians labored to inculcate in man and woman alike a deep sense of sin regarding the pleasures of the flesh. If religious taboos were insufficiently inhibiting, more practical fears were at hand: the fear of impregnating or of becoming pregnant, the fear of contracting a venereal disease, the fear of losing a spouse's devotion and of earning the disapproval of one's friends and the condemnation of society.

Departures from convention were never uncommon, human powers of self-discipline being what they are, but on the whole deviations from conformity have fared rather badly. Romantic love and other cultural variants have influenced people's attitudes from time to time and from place to place. But not at any time or in any place, until modern times, has there ever existed for very long any widespread belief that a stable society of responsible citizens could be maintained without at the same time maintaining the classical social institutions of marriage and the family. They clamped undeniable restrictions on individual freedom (or at least on individual license), but they also served the individual's essential needs. For a man, it served his need for sex, his need for a mate who would provide progeny to carry on his name, his need for status, his need to be needed, his need for a physical and psychical base of operations. For a woman, it served her need for security against the vagaries of circumstance during her periods of maximum vulnerability— pregnancy and child-rearing—her need for a man and a mate to provide her with children, *her* need for status, her perhaps even greater need than a man's to be needed. It was never perfect, but it worked better than any other system men had been able to devise—and most of us have been raised in the belief that, taking the facts of life all in all, things would go on more or less the same way in perpetuity. It was possible to believe this—almost impossible to believe otherwise—because there was no reason to doubt that the facts of human procreation would also remain essentially unchanged in perpetuity.

But in the sciences, even in the life sciences, perpetuity has a way of turning out to be not so perpetual after all. Birth control is certainly a phrase in common parlance, a matter of heated public debate and widespread private practice. Usually birth control is thought of only as contraception—through preventing ovulation (the descent of the egg from the ovary into the womb) or preventing fertilization (by killing the sperm, for example, or blocking its route to the egg). But birth control can mean, as well, *inducing* conception—e.g., by coaxing the egg, chemically, to enter the womb, by physically bringing sperm and egg together to insure (or at least increase the probability of) conception.

The possibilities for the control of birth—either to block conception or to stimulate it—are better understood by scientists than ever before. Information is being more efficiently disseminated to more and more people, and the barriers to its acceptance are noticeably diminishing everywhere. In addition to the variety of mechanical, chemical, and surgical techniques that already exist for the prevention of conception, scientists predict the reasonably early development of safer and longer-term contraceptives—pills, injections, or implantations that will last for weeks, months, or even years when desired; not to mention the further development of contraceptive drugs which may be taken *after* intercourse.* There also exist, only on an experimental basis at the moment, drugs that men can take to prevent the formation of sperm cells. For the encouragement of conception, where other therapies have failed, the doctor can intervene directly by introducing sperm (either the husband's own, or sperm from an anonymous donor) into the wife (artificial insemination), or by implanting in the wife an egg taken from the tubes or uterus of another woman (artificial inovulation).

The further refinement of freezing techniques will permit the establishment of sperm banks and egg banks. Long-term storage would mean that proximity in space and time of donor, recipient, and middleman (doctor) would no longer be required. Beyond artificially assisted fertilization, there could be (and has been) fertilization *in vitro*—the egg prefertilized before implanting in the wife. Or, going beyond that, it might well be possible to achieve what Dr. Petrucci claims to have begun—the growth of a baby *in vitro* with the protecting and nourishing presence of a human mother nowhere in evidence.

There are yet other variations to be played upon the theme of human procreation, variations which nature has already played spontaneously, at least with the lower orders of animals, and which biologists can now duplicate in the laboratory. There is, for one, the phenomenon of parthenogenesis, or virgin birth—or, in Dr. Rostand's phrase, "solitary generation." In mammalian reproduction, the organism normally gets half its chro-

---

* Because these substances do not prevent the sperm from penetrating and fertilizing the ovum—the classic definition of conception—they are not strictly contraceptives. What they do is prevent the newly-fertilized egg from implanting itself in the uterus. Since the interference occurs *after* conception, some hold that such practice constitutes abortion. A way around this impasse has been suggested by Dr. A. S. Parkes of Cambridge: Equate conception with the time of implantation rather than the time of fertilization—a difference of only a few days.

mosomes from the mother's egg, and half from the father's sperm.* But it is possible—on rare occasions in nature, as often as desired in the laboratory—for the egg to double its chromosomes. This gives it its full complement of chromosomes from the mother alone, making any contribution from the father unnecessary. Parthenogenesis is a common and normal occurrence in certain rotifers, insects, and crustaceans. Some species, such as the bag moth, *Solenobia triquetrella,* come in two separate and distinct races, one parthenogenetic and one strictly bisexual. By stimulating artificially (mechanically, chemically, or by the application of either cold or heat) the unfertilized eggs of sea urchins, frogs, and rabbits, scientists have been able to induce them to double their chromosomes parthenogenetically.** Spontaneous parthenogenesis in turkeys has been studied for years by Dr. M. W. Olsen of the U. S. Department of Agriculture. He and his colleagues raised a special breed of turkey with this parthenogenic tendency. "From the time of sex identification until maturity," says Dr. Olsen, in one of his reports, "the virgin females were reared in wire-enclosed pens to insure against any possible contact with sexually mature males. At the age of 30–36 weeks, the females were transferred to laying houses." Over a nine-year period many thousands of

---

* In the terminology of genetics, a cell containing a complete set of chromosomes in its nucleus is called a *diploid* cell. All the body's cells are diploid except the sex cells or *gametes*—which include both sperm and ovum. Gametes contain only half the required chromosomes, hence are *haploid*. When two gametes join, as when a sperm penetrates an ovum, the resulting fertilized cell becomes diploid. In parthenogenesis, an ovum somehow doubles its supply of chromosomes, thus converting itself from haploid to diploid without the intervention of a male sperm.

** Summarizing research in parthenogenesis at a 1952 meeting in Belfast, Dr. A. D. Peacock said: "Since Loeb's success [inducing parthenogenesis in sea urchins] there have been exploited at least 371 methods—I have counted them—45 physical, 93 chemical, 64 biological and 169 combined, not to speak of variations innumerable."

eggs were laid parthenogenetically. Most of them did not develop all the way. In the four years between 1956 and 1960 "a total of 67 embryos, all males, have survived to hatching, a few reaching maturity. Three males produced spermatozoa and one sired offspring."

Another possible source of virgin birth is said to be the hermaphrodite—an individual whose development in the embryo somehow went awry so that *neither* a definite male nor a definite female resulted, but rather a composite of both. In some species, hermaphroditism—the development of both male and female organs—is perfectly normal. In either case, an individual able to produce both sperm and ova might also be able to fertilize him-or-herself. Unlike the simple doubling of chromosomes, where no sexual activity is involved, this would in fact be a sexual union—the difference being that both the male and the female cells would have been supplied by the same individual. It is known that hermaphroditic animals—among them the earthworm and the sea hare—can achieve this kind of sexual self-fertilization. Is it possible in human beings?

Virtually every authority thinks not, because sperm and ova would almost certainly have to be produced alternately or successively rather than simultaneously. Yet there was a curious case reported by the late Dr. Walter Timme of New York, before the American Neurological Association, of a sixteen-year-old Arkansas girl who underwent surgery to have an ovarian tumor removed. The operation revealed that, in tissues adjacent to one another, there were signs of live ova as well as live sperm. "With the ova and the sperm in juxtaposition," said Dr. Timme, "there was a great possibility that they could combine and make a human being." He later averred that "there is absolutely no doubt whatsoever that this girl was a virgin. It was a true case of where the physical setup was possible for a virgin birth."

What about the possibility that the virgin birth of human

babies might come about through the doubling of the chromosomes—the kind of spontaneous parthenogenesis that seems to occur in Dr. Olsen's turkeys, for instance, and which the great Dr. Jacques Loeb was able to induce in sea urchins, and the late Dr. Gregory Pincus in rabbits?

A great deal of interest in this question was stirred in the years 1955 and 1956 in Great Britain by Dr. Helen Spurway, the wife of Dr. Haldane and, in her own right, a well-known lecturer in eugenics at London's University College. Dr. Spurway gave a talk called "Virgin Births," in which she suggested that, based on laboratory experiments with animals, it seemed likely that on rare occasions—say, once in 1.6 million pregnancies—there *would* be a spontaneous doubling of the chromosomes in the human female, and that occasional virgin births had probably already taken place, and might take place anytime, anywhere. The British medical journal *Lancet* did not laugh off the idea—though it did admit that "the possibility that a woman might become pregnant without at least one spermatozoon having entered the uterus is not one which the 'reasonable man' would lightly entertain," and, further, that "scientific opinion for several centuries has sided with the reasonable man."

If parthenogenesis in human females "does occur," said *Lancet,* "it is extremely rare. Suppose that its frequency were to be only one-half that for the birth of sextuplets—another rare event which undoubtedly does occur (with a frequency of about $1:80^5$ pregnancies or less).* Most of these cases would occur in women with a history of recent intercourse, and would be missed. In the few which might remain, the chance of any woman successfully asserting the facts in face of an evident pregnancy would be small enough to make her keep her own counsel."

---

* Unless my arithmetic is even rustier than I believe it to be, *Lancet* is postulating one virgin birth out of better than three billion pregnancies.

Both *Lancet* and Dr. Spurway suggested methods by which the claims of a parthenogenetic mother could be proved one way or the other. Blood comparisons of mother and daughter* as well as other scientific tests were described. Most crucial would be a skin graft from child to mother that would successfully "take" on a long-term basis. The immunological barrier to grafts, discussed in Part I, is such that only in the case of identical twins—where both individuals come from the same egg, thus have the same genetic characteristics—is it nonexistent. The same kind of rejection-defying graft capability in mother and child would seem to be proof that the child had come from an egg having no genetic characteristics but the mother's. (The positive results of this kind of skin-grafting test in turkeys has been considered strong corroborative proof of the offspring's true parthenogenetic origin.)

The British newspapers had a field day. The Manchester *Guardian* called it all "life *without* father." The tabloid *Sunday Pictorial,* enchanted by the whole idea, was not content merely with reporting its pros and cons. Its editors decided that if Dr. Spurway was correct there must be women in England even now who had given birth parthenogenetically and called upon any such women to come forward and declare themselves. Nineteen did, and Dr. Stanley Balfour-Lynn of Queen Charlotte's Hospital—supported by a battery of distinguished consultants—undertook to put them through the necessary tests.

There were some immediate dropouts when it became clear that many mothers had not properly understood what parthenogenesis meant. Of those that were examined, there finally

---

* The offspring of any such virgin birth would be female, though in some very rare instances, if the chromosomal information were in some way garbled, a male could conceivably result—but only a defective and abnormal male, unable to reproduce. Only the male sperm cells carry the Y chromosome, without which the making of a *normal* male is impossible.

remained one mother and daughter who met every single test. But there were still scientific skeptics, among them Haldane, who said the tests still fell short of constituting absolute proof that the mother had parthenogenecally created her daughter. Ultimately, the best Dr. Balfour-Lynn could do was to report: "In such a case as this, rigorous proof is impossible, but it remains that all the evidence obtained from serological and special tests is consistent with what would be expected in a case of parthenogenesis."

The case must be looked upon, then, as still in dispute. But some further evidence has been adduced in the laboratory. As long ago as the late 1930's, Dr. Stanley P. Riemann of the Lankenau Hospital in Philadelphia reported successfully inducing self-fertilization in a human egg by pricking it with a glass needle. The egg, though it released the "polar bodies" associated with a fertilized egg, only lived eight hours and never did actually undergo cell division. More impressive, though, are the more recent observations of Dr. Shettles. In studies of four hundred human ova, he found that three of these eggs had "undergone cleavage *in vivo* within the intact follicle, without any possible contact with spermatozoa." If three out of four hundred eggs had undergone cleavage—i.e., begun to divide—each in its protected little crypt, the follicle, before ever descending into the fallopian tubes or the uterus, it would seem that the eggs spontaneously underwent parthenogenesis, doubling their chromosomes, right there in the follicle. This could mean that parthenogenesis happens with considerably greater frequency than even Dr. Spurway guessed. And it would follow that virgin births have indeed occurred in human females, and that occasionally some young lady, puzzled and disbelieved on all sides, has become pregnant without ever knowing how it happened.

When parthenogenesis does take place, all the child's genetic traits are of course maternal, and there is only one true parent.

But a bit of microsurgery could easily make the *father* the one true genetic parent. The nucleus of the egg could be removed altogether, thus eliminating the mother's chromosomes, and replaced by a new nucleus containing a double dose of paternal chromosomes. This would constitute artificial androgenesis, and the resulting child would have no genetic mother at all.

Among other controls bestowed by medical science, especially through hormone therapy, will be increased powers to postpone or change the menstrual cycle at will, to abolish the pains and troubles of the menopause, and to strengthen masculinity or femininity when sex characteristics begin to flag.* Yet another will be the power to determine in advance the sex of one's offspring.

There are two kinds of sperm—one that produces males (androsperm), the other females (gynosperm). The kind that produces males, says Dr. Shettles, is somewhat lighter and speedier. This may be why slightly more boy babies are born than girl babies—though by the end of life there are more females because the males tend to die off a bit earlier. In the United States the greatest imbalance occurs between the ages of sixty-five and sixty-nine, when females outnumber males, 119 to 100.

Is there any way to separate the androsperm from the gynosperm to guarantee in advance the sex of a child? The answer would be simple if an evolutionary forecast made by Dr. Emil Witschi before he retired from the State University of Iowa had

---

* Many of these hormonal capabilities are already at least partially on hand and in use. In some cases hormone therapy is used to change the sexual characteristics of true biological males or females because they completely identify themselves, psychically, with the opposite sex. In extreme instances, drastic surgery has been performed to transform the person physically—and even legally—into a member of the opposite sex. Details of the therapy and its results are described by Dr. Harry Benjamin in *The Transsexual Phenomenon*.

already come to pass. Having studied man's evolution to date, Dr. Witschi predicted as the next logical step "a complete separation of the male line into gynosperm producing individuals and androsperm producing individuals." With two distinct breeds of men available, the woman could pick the sex of her child by picking her man. (Dr. Witschi did not discuss the possible marital complications of parents who wanted both sons and daughters.) Meanwhile a number of scientists, taking advantage of the slight differences—in weight, for instance—between androsperm and gynosperm, have claimed success in separating out the two types of sperm by electrophoresis, centrifugation, and sedimentation. By artificially inseminating animals with the separated sperm, they have managed to get a significantly higher proportion of the desired sex.* The results of these experiments have not been universally accepted. Dr. A. S. Parkes of Cambridge remains one of those still doubtful. But he says, in *Sex, Science and Society,* "I will have no doubt that some time, somewhere, such separation will be effected, either as the result of advances in relevant knowledge or by a stroke of inspiration. It may even be effected, as was the freezing of bull sperm, by people who in their innocence are unaware that what they are doing is well known to be impossible." The time may not be too far off, then, when expectant human parents can be guaranteed a boy or a girl by these methods or even better ones. Dr. Paul R. Ehrlich of Stanford believes this kind of sex choice is less than twenty years off.**

---

* Dr. Sophia J. Kleegman of New York University has been using her knowledge of the timing of ovulation to predict the sex of babies with 80 percent success. If insemination occurs 36 to 48 hours before ovulation, a girl is more likely, but if insemination takes place just at the time of ovulation, or very close to it, the probabilities favor a boy baby.

** Dr. Amitai Etzioni of Columbia University predicted, in *Science,* that it could come about in "five years from now or sooner." In Australia, Dr. Charles Birch of Sydney University was quoted as forecasting

Even after conception sex changes and reversals can still take place. Dr. Rostand has observed that the application of hormones at an early enough embryonic stage "upsets and reverses completely the processes of sexual differentiation. A male creature will be born from an egg that was originally determined as female" and vice versa. A number of scientists including Dr. Witschi and Dr. Carroll A. Pfeiffer of Yale have been able to bring about sex reversal in lower orders of animals in just this fashion. And Dr. Keith Moore of the University of Winnipeg has reported that on occasion the same kind of sexual transformation occurs quite spontaneously in the embryo.

Clearly, the "facts of life" are mutating, and as they undergo these artificially induced mutations, there will in turn be a chain reaction of mutations in our social attitudes and institutions.

People tend to overestimate the probable resistance to new practices—*in vitro* embryology, for instance. Not that we know how soon this particular piece of bioengineering will actually be available to us. Before a full-term grown-in-glass baby can be brought forth into the world, a number of formidable technical obstacles will have to be overcome. Not the least of these will be the necessity for duplicating the complex and crucial functions of the placenta throughout fetal development. A number of research teams in the U.S., England and the

---

that husbands may one day decide the sex of their offspring by taking pills in advance. In Washington, Dr. E. James Leiberman of the National Institutes of Health suggested that one day women might wear "a special diaphragm that would let through only the sperm that carry, say, the male sex, and holds back those that carry the female sex." These latter notions seem unlikely to me—which only illustrates once again that scientists are often willing to speculate more daringly than journalists.

U.S.S.R. are studying these functions and seeking to simulate them artificially. However long it takes, it is a fair assumption that none of the obstacles will turn out to be insurmountable. And when they are all surmounted, someone somewhere is going to produce *in vitro* offspring. "If I can carry a baby all the way through to birth *in vitro*," says an American scientist who wants his anonymity protected, "I certainly plan to do it—though, obviously, I'm not going to succeed on the first attempt, or even the twentieth." He suspects that the public indignation here might be as loud as it was in Italy.

Indignation or not, once the first success has been achieved, it will not be too long before some people in special circumstances start raising babies in this fashion. Would anyone consent to something so "unnatural"? Think back to a time when word was first getting around that something called artificial insemination was being practiced in some daring animal-breeding experiments. Imagine a group of people at a nineteenth-century dinner party, with someone describing what this technique entailed—and suggesting that it might some day be applied to human breeding. There would have been shock, disbelief, amusement, anything but serious conviction that this might really happen. What man would ever donate or sell his sperm to impregnate some strange woman at a distance? And what kind of woman would it be who would allow herself to be thus impregnated? And what self-respecting husband would ever consider such an outlandish route to fatherhood? And what ethical doctor would participate in such an act? Yet, now that the novelty of the idea has long since waned, it is an everyday occurrence for sperm donors, childless parents, and responsible physicians to collaborate in artificial insemination.

So would it be with *in vitro* babies. Besides, the sheer medical advantages may be too great to resist. Medical scientists have only lately begun to explore the field of prenatal medicine (also called fetology, fetal medicine, or neonatology)—the

117

study and treatment of the embryo or fetus* while still in the womb. As Dr. Robert Aldrich of the University of Washington School of Medicine points out, most of the congenital defects that afflict newborn babies occur during this critical period of development, at a time when they are shut off from any observation, diagnosis, or therapy. As the scientific exploration of prenativity speeds up, doctors will be increasingly able to spot and correct defects and add benefits to a baby while still in its malleable embryonic state. After birth is often too late.

Fetuses can already be observed, up to a point, by various methods: X-rays, radio-opaque dyes, infrared thermography, ultrasonics, monitoring microphones. Nor is it any longer a rarity, since the daring procedure was first introduced by Dr. A. William Liley in New Zealand, to hear that an Rh-negative fetus has been completely transfused inside its mother. Dr. Jacobson and others have perfected techniques for analyzing the amniotic fluid which surrounds the fetus to the point where they can diagnose upwards of twenty fetal diseases.

Even with the most advanced techniques, however, there are definite limits to what can be accomplished while an embryo is inside the womb. But if it were *in vitro,* visible and accessible, incomparably more could be done. One need only look at a tiny fetus through the viewing porthole of Dr. Robert Goodlin's steel experimental chamber at Stanford—or at a similar primate fetus, in this case still attached to its mother via the umbilical cord, at the University of Nevada—to see how convenient observation and treatment could be. Faults could be detected much more readily, and some could be entirely avoided merely by protecting the embryo from exposure to

---

* The term *embryo* is applied loosely to the entire prenatal state, but technically an *embryo*—from the Greek word meaning "to swell"—becomes a *fetus*—from the Latin word meaning "young one"—at the age of eight weeks. Key to the switchover is the replacement of cartilage by the first true bone cells.

harmful drugs and viruses—e.g., German measles—to which it would have been vulnerable in a real mother's womb. On the other hand, helpful drugs and hormones could be applied and minor surgery performed with much greater ease and control. Dr. Medawar won his Nobel Prize for experiments with mice which offer hope that treatment during a critical phase of embryonic growth might provide a child with later immunological tolerance to transplanted organs. *In vitro* treatments would be far surer and simpler.

The crucial early stages of embryonic development might be hardest to achieve *in vitro*. But some of the benefits might be applied even before all the problems are solved. "An intermediary solution of the problem of pregnancy is, indeed, conceivable," says Dr. Rostand. "Delivery could be stimulated artificially and the embryo placed in culture at the age of two or three months: in short, a woman would reproduce like a kangaroo.

"If ever partial or total 'test-tube pregnancy' came to be applied to our species, various operations would become possible, resulting in a more or less profound modification of the human being in course of formation." At the Oregon Regional Primate Center, monkey fetuses have been removed from the womb, had surgery performed on them and electrodes attached to them for electrocardiograph readings before being put back into their mothers and proceeding to apparently normal deliveries.

"It would be no more than a game," says Rostand, "for the 'man-farming biologist' to change the subject's sex, the colour of its eyes, the general proportions of body and limbs, and perhaps the facial features.

"Is it very rash to imagine that, in that case, it would be possible to increase the number of brain cells in the human member? A young embryo has already in the cerebral cortex the 9,000,000,000 pyramidal cells which will condition its mental activity during the whole of its life. This number, which is reached by geometrical progression or simple doubling, after 33 divisions of each cell (2, 4, 8, 16, 32 and so on), could in turn be doubled if we succeeded in causing just one more division—the thirty-fourth." Making the brain or head much larger had earlier been pooh-poohed on the grounds that normal birth would be rendered impossible. But Rostand points out that this objection would be irrelevant if the child were to be produced in an artificial womb—whose opening could presumably be made as large as one wished. Remember Huxley's Predestinators who in *Brave New World* imparted the desired characteristics to developing embryos? One wonders who would perform this function for us in our brave new society.

Apart from medical uses, there is the great convenience for the mother of not having to undergo pregnancy at all, of skipping the morning sickness, the heavy step, the kicking fetus, the labor pains. Of course, there will remain staunch old-fashioned types who consider this more a deprivation than a convenience. Many people feel that it would require a thorough-going revolution in ethics and morals to make this whole idea acceptable. Yet sex as recreation as opposed to sex as procreation has long since gained popular acceptance. And how many women have ever turned down labor-saving devices?

Who would have predicted that so many millions of women would be so eager to interfere with their natural cycle of ovulation by dosing themselves with birth-control pills every day? If they could safely imitate the habits of, say, the lady ostrich, or the mama penguin of Antarctica, who deposit their eggs and then go off to run and play and forget them, who is to say that they would not jump at the opportunity?

We need not wait for the onset of *in vitro* technology to witness bizarre happenings in the field of reproduction. Not long ago Dr. E. S. E. Hafez, an Egyptian-born experimental biologist now at Washington State University, told me he had just commissioned a scientist friend from Germany to bring him a hundred head of sheep. The entire herd was to be delivered to him in a neat package he could easily carry in one hand. The package would be a ventilated box. Inside the box would be a small female rabbit. Inside the rabbit would be a hundred incipient rams and ewes, all of them embryos only a few days old, growing there as if in their natural mother.

This means of transporting livestock is only one unusual aspect of a novel brand of animal husbandry of which Hafez is one of the most active and enthusiastic practitioners—though certainly not the only one, nor the first. His specialty is producing multiple births, in such enormous litters that he considers mere quintuplets a failure. Through the ingenious use of hormones and other animals that serve as foster mothers, he routinely produces centuplets.

Normally a cow's ovary releases one egg at a time, just as the ovary of a woman does. As a result, though a cow is born with thousands of incipient eggs in its ovaries, it can give birth to no more than ten or twelve calves in its lifetime. We have already seen that ovulation can be induced by the administration of hormones when the eggs fail to come down as they normally should. But it is also possible to induce *superovulation*—making the animal's ovaries release ripe eggs in quantity far beyond anything that might normally occur. Hafez, who learned the techniques from scientists in Great Britain, where much of the pioneering research was done,* uses two sets of

---

* The earliest experiments were done around the turn of the century by men ahead of their time—men like Dr. Walter Heape and Dr. F. H. A. Marshall. More recently, and in a more congenial climate, exciting work has been done at Cambridge's School of Agriculture by Drs. C. E. Adams, L. E. Rowson, G. L. Hunter and R. L. W. Averill. Important progress in this area of research has also been carried out at the Japanese Ministry of Agriculture.

121

hormones—one extracted from pregnant mares, the other from pregnant women—to superovulate his cows. And it works just as well with immature calves or with very old cows as with animals in the normal calf-bearing years. This same family of hormones has recently gained fairly wide use for the purpose of rendering barren women fertile, sometimes resulting in multiple births.

Newspapers in recent years have carried stories about sets of quintuplets born within short periods of time as far apart as New Zealand and Sweden. The quints were born through the administration of hormones to the mother. So were a rash of quadruplets, triplets, and twins. In these human cases the multiplicity was inadvertent. In Hafez's cows it is deliberate. Once his cows have undergone multiple ovulation, he goes through a procedure first successfully carried out by Dr. E. L. Willett and his associates at the University of Wisconsin and the nearby American Foundation for the Study of Genetics. Only he does it on a wholesale scale. He fertilizes as many as a hundred eggs at a time via artificial insemination, all in the same animal. Several days after the mass conception, he flushes out the tiny embryos. These can then be implanted directly into the uteruses of other cows, one or sometimes two to a cow. If no mishap occurs and if the receiving uterus has been properly prepared (with the hormones the cow would be secreting if it were naturally pregnant), the implanted calf will go through a normal gestation period and be born as if it were the foster mother's own—but possessing the genetic qualities of its true parents. As an interim measure, the embryos can be transplanted into smaller animals, such as the rabbit which Hafez's friend was going to bring from Germany. In their temporary abodes, the embryos continue to grow normally for up to fourteen days before they have to be transplanted into other cows.

Hafez has visions of full-scale commercial operations where whole herds of superb, pedigreed cattle could be flown cheaply and easily across the oceans—say, from the United States to Brazil—inside a single rabbit.* If Hafez's plans work out, animals of the imported breed could be carried in the wombs of any scrub cows, acquiring from their foster mothers immunity to the local diseases while retaining all the prize-winning hereditary characteristics of their genetic parents. Meanwhile their original mother grazes contentedly back on the farm, ready to produce the next batch of eggs on hormonal demand.

Hafez sees no reason why these techniques would not work just as well with people. As he unfolds his plans, the corners of his mouth crinkle downward in wry amusement and an antic twinkle lights his eye, and one is again uneasily aware of the fuzzy borderline that seperates the "mad scientists" of fiction from the real ones whose sanity is unquestioned. Hafez is head of the Department of Animal Sciences at Washington State University, and his research has been generously supported by the National Institutes of Health.

Hafez has in his laboratory a set of symbolic vials bearing labels like "cattle," "sheep," "swine," "rabbits," "man." Placing these in the palm of his hand, he says, "This could be the barnyard of the future—complete with farmer." Fertilized egg cells could be shipped anywhere in the world in cold storage in such containers, each vial color-coded to identify the species it contains.

When man begins to colonize the planets, Hafez is certain that his techniques are the only sensible means to employ. "When you consider how much it costs in fuel to lift every

* Transportation of sheep embryos via rabbit has already been carried out—from England to South Africa. The first results, in 1961, were healthy, pure white Leicester lambs out of blackfaced South African ewes. Dr. Parkes, describes the rabbits as "handy thermostated containers, cheaper to send by air than sheep and less likely to arouse the interest of veterinary officers at the port of entry."

pound off the launch pad," he argues, "why send fullgrown animals, men, and women aboard spaceships? Instead, why not ship tiny embryos, in the care of a competent biologist who could grow them into people, cows, pigs, chickens, horses— anything we wanted—after they get there? After all, we work hard to miniaturize the spacecraft's components. Why not the passengers?"

Hafez has been working out the details of a plan for populating Mars. So convinced is he of its feasibility that he fully expects to be awarded a government contract to help his Martian program along.

Though dreaming of planetary conquests, Hafez has notions for the earth too. One of them is first cousin to a scheme which, when it was presented a few years ago on the Broadway stage, served as pure comedy. In James Costigan's play, *Baby Want a Kiss,* Edward says to his friend Emil, whose success he adulates:

> "The only wonder to me is that somebody doesn't put you up in packages and sell you in all the supermarkets and discount stores."
>
> Emil replies, "You think I haven't thought of that? Not just me, but 10 or 15 top celebrities like me."
>
> EDWARD: "In packages?"
>
> EMIL: "Frozen."
>
> EDWARD: "Frozen?"

In a moment Emil confides: "It's the coming thing. I've put a fortune into researching it. Now listen to this: We call it Celebrity Seed . . ." Then, imitating a TV announcer: "Ladies why settle for second best? Now you can have a baby by your favorite actor, singer, dancer, band-leader, TV announcer, TV panelist, cafe socialite, colum-nist, golfer, baseball star, congressman, senator or, in certain isolated instances, dress designer . . . Celebrity Seed is on sale, without a prescription, at your corner drugstore.

For the corner drugstore, Hafez substitutes a new kind of medical commissary. He speculates that within fifty years it could be possible for a housewife—armed perhaps with a doctor's prescription, or whatever kind of license the system required, since it would be no ordinary commercial purchase—to walk into such a commissary, look down a row of packets not unlike flower seed packages, and pick her baby by label. Each packet would contain a frozen one-day-old embryo, and the label would tell the shopper whatever details could be predicted about the probable characteristics of the child. It would also offer assurance of freedom from genetic defects such as sickle-cell anemia and mongolism. After making her selection, the lady could take the packet to her doctor and have her newly-purchased prefab embryo implanted in herself, where it would grow for nine months, like any baby of her own.

If the prospective mother is not the type who delights in the joys of pregnancy, perhaps *in vitro* technology will be so far advanced by then that, instead of heading for the doctor's office, she can just drop her packeted baby off at the bottling works and pick it up when ready. If she is a sentimental sort, she can come by now and then to see how the child is getting on. (When it does become possible to *see* a baby growing, practically from the day of conception, we may switch to the sensible old Chinese custom of counting that day as the child's true birthday—with a consequent effect, no doubt, on our attitudes toward abortion. Many readers of *Life* who saw Lennart Nilsson's marvelous photographs of fetuses in their sacs, especially in the later stages of development, wrote in to say that they could never again think of those *babies* as disposable *things*. Such sentiment might well increase as fetuses become

125

visible from the outset. And if the day of conception were to become a person's official birth date, then the act of aborting a fetus would be ending the life of a baby of a given age.)

Growing babies *in vitro* and producing centuplets by super-ovulation may seem to be daringly sophisticated departures from nature's leisurely and secretive routes to procreation. But as time goes on, even the methods employed by Hafez and Petrucci, and all subsequent refinements of them, are bound to be looked upon as relatively crude ways to raise human beings —or whatever the classification of such beings turns out to be in the taxonomy of the twenty-first century. More efficient methods have already been sketchily described, methods which altogether bypass the awkwardness and uncertainty of mating sperm with ovum. Without rehashing the endless debate as to whether or not life will ever be created from inanimate chemicals in the laboratory, some hair-raising projections can be made from work already done and in progress.

As far back as 1891 the German biologist Dr. Hans Driesch had dramatically shown that the original fertilized egg is not the only cell that can develop into a whole organism. He took a sea urchin embryo in its very incipient stages, when the original cell had divided twice (first into two cells, then each into another two, making four in all), was able to shake the four cells apart—and each of them grew into an entire sea urchin. But later scientists were able to interfere in the embryonic process in a much more detailed and sophisticated manner.

At the University of Indiana during the 1950's, for instance, Dr. Robert Briggs and Dr. Thomas King did some delicate surgery on the nuclei of frogs' eggs. Operating on a frog embryo, they totally removed the nucleus from the original fertilized egg cell. The nucleus contained all the frog's genetic instructions on how to continue growing. Without these instructions, any organism must perish. But in this case Briggs and King replaced the missing nucleus with a nucleus from another of the embryo's many cells—and the embryo kept right on growing as if nothing had happened. Successful repetitions gave assurance that the phenomenon was not a one-time freak occurrence. The meaning was clear: The nuclei of the other cells must have also contained the entire manual of instructions.

This, on the face of it, is not too surprising. We know that a cell, when it divides, does so in a way that provides each of the two new cells with a whole, perfect nucleus—including all the genetic information. But the trick performed by Briggs and King works only during the first few days of the embryo's life. After that, transplanted nuclei will no longer do the job. The genetic information begins to be used and inhibited selectively. Each cell employs only the specific instructions it needs at each step along the way to differentiation—to becoming, say, muscle or nerve tissue. It also knows when to stop growing as well as how to manufacture selected enzymes and antibodies under selected sets of circumstances.

These biological timing mechanisms are not the same for all species, however. Briggs and King carried out their provocative experiments on a frog called *Rana pipiens*. But, switching to the African clawed toad, *Xenopus laevis,* Dr. J. B. Gurdon and his associates at Oxford were able to produce even more impressive results. In *Xenopus,* the nuclei of the embryonic cells retained their generative capacities at much later stages of development. Even cells from well-differentiated parts of the embryo—the neural fold and tailbud, for example—could be substituted successfully for the egg nucleus. In one instance, an intestinal cell taken from an almost fully developed tadpole and transplanted to an egg resulted in a normal toad. But even using *Xenopus,* there were limits beyond which the researchers could not go. Like Briggs and King, they discovered that a cell's generative capacities declined with age.

In a fullgrown, mature organism, every cell—no matter how

different it might be in form and function from other body cells—still possesses, if it has developed normally, all the genetic data transmitted by the first fertilized egg cell. But most of the data has become permanently masked, since the body under normal circumstances is unlikely ever again to have use for it. As we now know, the masking appears to be chemical in nature—and if this is so, there is no theoretical reason why a clever counter-chemistry might not be devised to unmask it. If cells can be thus induced to shed their inhibitions, making accessible once more all the genetic data in the nucleus—endowing it with what Rostand calls "seminal virtue"—should it not then become possible to grow the organism all over again from any cell taken from anywhere in the body? From such speculations grew the notion of raising people in tissue culture.* Remember what happened when an egg was "bokanovskified" in *Brave New World?* "Making 96 human beings where only one grew before. Progress."

Farfetched? Certainly. Yet, at Cornell University, Dr. Frederick C. Steward has been achieving exactly this sort of asexual reproduction with the lowly carrot and tobacco plant. After years of research, Dr. Steward has reached a point where he can routinely take a single cell from adult plant tissue, treat it chemically to make the original genetic instructions available once more, then grow from this cell an entire, healthy plant capable of bearing seeds and reproducing itself normally. One can look at a piece of glassware in Steward's laboratory and see thousands of tiny carrots beginning to sprout from what were thousands of individual carrot cells chipped randomly from various parts of the adult carrot. (It still costs more to raise them this way than agriculturally, however.)

It is of course a very long way from carrots and tobacco

---

* The technical term for this kind of cell growth in tissue culture— "cloning"—has lately been finding its way into the popular literature.

plants to animals and people. Dr. Steward cautions that animal and human cells may behave very differently from plant cells. Here again, a series of breakthroughs are required to overcome the technical barriers that still lie ahead before man can master the incredibly complex electrochemical switching networks that govern cell growth and development. Meanwhile important transitional work in this direction is being done by Dr. James Bonner and his colleagues at Cal Tech. It was on the basis of work like Steward's and Bonner's that Rostand was moved to point with an air of triumph to scoffed-at words he had written many years earlier. "If a biologist," he had said at the time, "takes a tiny fragment of tissue from the freshly dead body, he could make from it one of those cultures which we know to be immortal, and there is no absolute reason why we should not imagine the perfected science of the future as remaking from such a culture a complete person, strictly identical to the one who furnished the principle."

Since Rostand wrote those lines the presumed immortality of tissue-culture cell lines has been called into question. As a result of work done at Philadelphia's Wistar Institute by two University of Pennsylvania investigators, Dr. Leonard Hayflick and Dr. Paul S. Moorhead, it turns out that only abnormal cell lines—such as the malignant HeLa strain—achieve a true immortality. Normal cell lines have a definite built-in life span and die out after a certain number of generations (i.e., cell divisions). As pointed out by Editor Joan Woollcott in *Medical Affairs,* "It is possible to arrest cell growth and to store normal human cells in a state of 'suspended animation' at sub-zero temperatures for apparently unlimited periods of time."

" 'We have stored such cells in liquid nitrogen at various generation levels, from 1 through 50, for as long as six years,' Dr. Hayflick reported.

"When these cells are reconstituted in tissue culture, they begin to divide again. They 'remember' what generation they had reached before storage, and they take off from that point. For example, cells stored at the 20-generation mark will divide for 30 more generations after they have been reactivated.

"By maintaining a sub-zero cell bank, with cultures labeled by generation, investigators can keep always on hand for research purposes a strain of known and tested normal human cells, which can be activated and expanded at will."

The new knowledge of the mortality of cell lines does not in any way rule out the kind of tissue-culture reproduction forecast by Rostand—and independently by Haldane. "This new technique of generation from the nucleus of a body cell,"* says Rostand, "would in theory enable us to create as many identical individuals as might be desired. A living creature would be printed in hundreds, in thousands of copies, *all of them real twins*. This would, in short, be *human propagation by cuttings,* capable of assuring the indefinite reproduction of the same individual—of a great man, for example!"**

To this kind of proposition, Dr. Theodosius Dobzhansky of Rockefeller University replies: "It can show no lack of respect for the greatness of men like Darwin, Galileo, and Beethoven, to name a few, to say that a world with many millions of Darwins, Galileos, or Beethovens may not be the best possible world." Dr. George Wald of Harvard likes to tell about one of his recurrent nightmares: He walks into Grand Central Station

---

* Could we, by this method—if it had already been perfected—have undone the assassination of, say, President Kennedy? But it would still have taken a lifetime of years to produce a new John F. Kennedy of the desired age.

** Remembering the possible deterioration of, or damage to, the nucleic acids inside the cell with aging, Rostand suggests the wisdom of putting aside in the deepfreeze, or in whatever kind of long-term storage becomes feasible, tissues of a person when he is *young,* thus increasing the probability of intact genetic material when the time comes to use it.

and sees eight Albert Einsteins buying eight copies of *The New York Times*.

People grown in this fashion in tissue culture might indeed be "real twins," as Rostand says, but would they really be identical? Would another human being grown from a cell of Albert Einstein turn out to be another Einstein? Even though the genetic information is the same, how much of Einstein's potential may have depended upon the environment in his mother's womb? And how much of his realized genius was due to his upbringing and to his unique, unduplicatable experience-in-history? It does seem likely that men and women with Einstein's genetic heritage would have a higher than average chance to possess great intellectual and creative possibilities, and therefore to contribute something worthwhile to society. But it is quite possible, that of a hundred or a thousand "real twins" of Albert Einstein grown in tissue culture, not one would have thought of the theory of relativity.

(Picture a young scientist whose father has been killed, growing himself a new father from one of his father's still-living cells. Who would now be whose father? Would the new baby call the scientist "Daddy"—or vice versa?)

It might of course be possible to raise a multiplicity of identical individuals from the very start, beginning with the fertilized egg, rather than waiting to take a cell from a grown man or woman and endowing it chemically with seminal virtue. Now and then a fertilized egg does split, on its own, into two separate fertilized eggs before it begins its normal development—and the result is identical twins. The armadillo egg behaves this way routinely; it splits four ways, and armadillos regularly give birth to identical quadruplets—performing naturally what Dr. Driesch got sea urchins to do artificially. The egg of a chalcid wasp can perform even more impressively; it can keep splitting until it gives birth to centuplets—but not like the artificially induced bovine centuplets of Dr. Hafez. His are all separate individuals, the products of a hundred separate eggs fertilized by a hundred separate spermatozoa. The chalcid wasp produces *identical* centuplets.

Even if we knew how to do this with human fertilized eggs, there would be a built-in handicap: there would be no way to judge, so early, how the individual was going to turn out, and thus whether or not he would be worth duplicating and reduplicating. This handicap aside, though, there is again no theoretical reason why, if we knew the mechanism of twinning, we could not teach any fertilized egg to turn itself into two, and to teach each of the resulting two to turn into two more, and so on—just as we might teach an ovum to double its chromosomes to become parthenogenetic.

Some animals, notably insects like the honeybee and the greenfly, can be parthenogenetic part of the time, and sexual the rest of the time, as it suits the seasons and circumstances. Humans could probably do the same, if they chose. But in the case of parthenogenesis, what advantages would there be from practicing it, unless one wanted only his (or only her) genetic characteristics imparted to the offspring, or unless one hated the opposite sex—or his own?

On the other hand, if either parent possessed a known genetic defect, he might well prefer to give his child only the genetic characteristics of the other parent. It is also quite possible that an unmarried young lady might want to make use of induced parthenogenesis to have a baby. In Sweden today, it is anything but uncommon for young unmarried women to have babies. When they do, they are called "Mrs.," they are entitled to all family allowances, and their children are considered legitimate. Though it is still looked upon as rather disgraceful to be an unwed mother in most countries, people are getting more relaxed about it than they used to be. So fear of disgrace would probably not be much of a deterrent to women who badly wanted children—and who preferred to have them nonsexually. Dr. Rostand feels certain that if induced parthenogenesis were ever to become available and reliable technique, at least some women would want to take advantage of it.

Societies can of course be conjured up in the imagination that would be virtually all male or all female. In an Amazonian culture where *females* have gained ascendancy and males are held in contempt, the race could be propagated in a number of ways: by keeping a few domesticated males in captivity like stud bulls, by drawing semen from frozen sperm banks, or simply by artificially doubling their own chromosomes. In Charles Eric Maine's *World Without Men*, one of the female scientists becomes a fugitive in order to keep an experimentally produced male baby from being destroyed on orders of the security council.

And, similarly, one can well imagine a male society. Today, homosexual behavior is growing more rampant and more blatant. In those ancient Greek societies where only love between men was deemed noble, women were often looked upon as an unavoidable nuisance necessary for continuing procreation. A modern misogynist society could now begin to draw on egg banks—and if its leaders objected to the female chromosomes in the eggs they could simply substitute their own. The practice of "auto-adultery," as that inexhaustible phrasemaker, Dr. Rostand, calls it, would have to satisfy them, at least, until the day people can start doing with themselves what Dr. Steward has been doing with carrots. Then they can raise people nonsexually, and neither sex would be necessary to the other for any phase of reproduction.

We have already imagined a future society ruled by a dictator with all these new techniques at his command. In our own century, which has seen whole peoples extirpated in the name of totalitarian ideologies—and where even the fictional

Utopias (George Orwell's *1984*, Ray Bradbury's *Fahrenheit 451*) are depressingly pessimistic—we entertain few doubts that unscrupulous opportunists and power-seekers will make use of whatever knowledge is at hand in any way that serves their ends. Even by using such relatively crude psychosocial devices as propaganda and brainwashing—but using them expertly and ruthlessly—monumental feats of evil have been perpetrated. So a decision to ignore these advances might be merely a decision to turn them over to the opportunists.

A future dictator could grow armies, or harems, of any size. He could produce latter-day myrmidons by the millions, and, turning them loose as suicide troops or saboteurs, endanger nations much larger than his own. He might insure his personal immortality as well as the perpetuity of his reign by arranging to have several versions of himself, at various stages of growth—one always ripe-and-ready to take over in the event of his death or assassination. Assassination would of course be unlikely in the midst of such universal loyalty. Yet, something could go wrong. It might not be feasible to have every single individual under such total control. Some tasks would still require some initiative. Someone would have to program the computers to program the people. Someone would have to run the dictator's people farm.

Suppose the scientist in charge decided to turn loose one of the heirs-apparent ahead of schedule? The dictator might then lose his job, and his head, prematurely. Meanwhile, what would the dictator decide to do with all the versions of himself he had no use for? Would he consider them threats, and have them executed? And if these new persons are simply new versions of the same old dictator, bearing the same identity, would this constitute multiple murder? Multiple suicide? Not that it would matter, since no one would be in a position to prosecute.

Long before the arrival of such an anti-Utopia of the future, might some enterprising captain of industry be tempted to grow his entire labor force *in vitro* or in culture? Such laborers would never strike, or even demand a new contract.

Can we learn to exercise control over our new controls, so that man may use scientific advance as it should be used—as a tool to serve human values in a kind of society that can still be called democratic?

If we need answers to deal with the dilemmas of sexless reproduction, we will need more complex answers, and we will need them more urgently, on whatever day we arrive at the ultimate in controlled procreation—at least, the "ultimate" we can now project: the production of beings whose specifications can be drawn in advance.

When this kind of biochemical sophistication has been attained, when man can write out detailed genetic messages of his own, his powers become truly godlike. Just as man has been able, through chemistry, to create a variety of synthetic materials which never existed in nature, so may he, through "genetic surgery," bring into being new plants, new foods— and species of creatures never before seen or imagined in the universe—beings that might be better adapted to survive on the surface of Jupiter, or on the bottom of the Atlantic Ocean.

Whether or not he chooses to exercise his new talents in those directions, man will presumably be able to write out any set of specifications he might desire for his ideal human being. This is one of the powers scientists have in mind when they talk about man controlling his own evolution. And who can find fault with creating ideal human beings? Is there anyone not in favor of emphasizing man's good qualities and eliminating the bad ones? The rub, of course, is that "good" and "bad" are words that are easier to say than to apply.

Most people would agree that it is good to reduce infant mortality, to make it possible for infertile parents to have children, to eradicate cancer and heart disease, to avoid heredi-

135

tary shortcomings from harelip to hemophilia. But consensus is not unanimity. Dissident voices are even now insisting sourly that most of our so-called advances actually militate against human progress by aggravating the population problem, thus assuring an overcrowded planet where more people will die of war and starvation. A good many others believe that medical advances may well result in the deterioration of the human race because so many people with hereditary defects, people who formerly would have died at an early age, are now kept alive to marry and pass on their defects.

"It is a depressing thought," writes Dr. Dobzhansky in *Heredity and the Nature of Man,* "that we are helping the ailing, the lame, the deformed, only to make our descendants more ailing, more lame and more deformed. Here, then, is a dilemma—if we enable the weak and the deformed to live and to propagate their kind, we face the prospect of a genetic twilight; but if we let them die or suffer when we can save or help them, we face the certainty of a moral twilight." No scientist advocates letting genetically defective individuals suffer unnecessarily, but there is a growing sentiment in favor of restricting their procreative prerogatives.

"There may well come a time," say George and Muriel Beadle in *The Language of Life,* "when even as humane a society as ours will find that people with inherited diseases constitute a social burden so great relative to our resources that we will be forced to limit what most of us consider one of our inalienable individual rights—the right to bear children without reference to the consequences for society. We already do this to a limited extent when we isolate certain categories of feeble-minded, insane, and criminal individuals—and thus prohibit their reproduction."

There is always someone to argue, too, that men with severe physical handicaps—Steinmetz, for example, or Toulouse-

Lautrec—often make invaluable contributions to human welfare; or to point out that the genetic endowment of Abraham Lincoln, who was afflicted with Marfan's Syndrome—a congenital ailment characterized by abnormally elongated fingers (and sometimes toes), unusual flexibility of the joints, a slight dislocation of the eye's crystalline lens, and perhaps hidden cardiac and spinal defects as well—might have ruled him out if strict eugenics had been practiced in his parents' day. "It is a question," says Thomas Mann, "of who is sick, who is insane, who is epileptic or paretic." Writing of Dostoevsky, who was epileptic, and Nietzsche, who was paretic, Mann points out that "in their cases the disease bears fruits that are more important and beneficial to life and its development than any medically approved normality." Achievements like theirs, Mann believes, "force us to reevaluate the concepts of 'disease' and 'health,' the relation of sickness and life; they teach us to be cautious in our approach to the idea of 'disease,' for we are too prone always to give it a biological minus sign." So it is quite possible that, in curing the world's ills, scientists would risk the loss of an occasional genius.

The most popular counterargument is that the intelligent control of heredity could produce many more geniuses by design than affliction has produced by chance, at the cost of untold suffering. Besides, despite Mann's admonitions, there is little doubt that most prospective parents would choose to have a child of sound mind and body than a flawed and unhappy genius. Most people would also choose—at least for themselves and those they love—the benefits of health and longevity, and worry about the population problem later.

Among the powers which most people would probably welcome would be those bestowed by hormone therapy—all those powers to suppress or stimulate ovulation for birth control purposes, to postpone or change the menstrual cycle, to control or abolish menopause, to augment either masculinity or femininity, to control growth so a child would not be too small or too tall. Equally welcome would be the forthcoming power to decide, in advance, the sex of one's offspring. There would certainly be medical advantages to this last-mentioned capability. "For example," says Dr. Jacobson, "suppose a man knows that he carries a recessive gene for hemophilia, though he does not have the disease himself. He knows, further, that he cannot pass on this gene to his sons, but only to his daughters—who, though they themselves will not have hemophilia, will pass it on to their children. Such a man, by deciding to have nothing but sons, could put an end to hemophilia in his descendants at a single stroke."

The elimination of sex-linked hereditary diseases is one of the goals motivating the work of Dr. Robert Edwards and Mr. Richard Gardner at Cambridge University. Unimpressed by the attempts to date to control sex in advance by separating sperm, Edwards and Gardner have pursued another route: Through the study of rabbit embryos, they believe they can identify the sex of an embryo by examining the cells for their sex chromatin. Moreover, they can do it in the embryo's very earliest stages, before it becomes embedded in the wall of the womb. The inspection requires the removal of the embryo, but it can be returned to its mother, or to the uterus of a foster mother. There is some risk of damaging the embryonic cells in the process, but even at this early stage of experimentation Edwards and Gardner have been able to bring eighteen of their first twenty rabbit fetuses through to full-term development and normal birth—and, in all eighteen cases, the sex turned out as predicted.

They realize that, if the aim is to assure a child of a given sex, then there should be more than one embryo to choose from. They suggest superovulation as one means of bringing

138

down and fertilizing many eggs at once—then, after examination, putting back only the embryo selected.*

Among genetic disorders that are sex-linked, Edwards and Gardner list, besides hemophilia, "one form of muscular dystrophy and several enzyme-deficiency diseases, all of which are many times more common in males than in females." "The elimination of these disorders in one generation, by a judicious choice of the sex of the offspring," write Edwards and Gardner in *The New Scientist,* "would not only be of direct benefit to that generation, but would benefit the race for generations to come. More immediately, the ability to determine the sex of domestic animals would be of enormous practical importance to farmers."

The usual reason for determining in advance the sex of human babies, however, would simply be the preference of the mother or father. But this kind of parental prerogative could work out to the disadvantage of the population at large. If people could be counted on to prefer well-balanced families of boys and girls, there would be no problem. But if this turned out not to be the case, the balance of the sexes could easily be destroyed. If there were a vogue, as there has been at some times and in some places, for either boys or girls, the result could be an acute shortage of one or the other. The further result would be a plethora of bachelors or of old maids—or, if things got desperate, a switch to polygamy or polyandry.

Most of the medical and biological tampering so far discussed comes under the heading of *negative eugenics*—that is, eugenics which concerns itself purely with getting rid of diseases

---

* This would raise—as in the case of the "morning-after" pill—questions of abortion. But here again Dr. Parkes' solution might work: to count conception as occurring at the time of embedding in the womb rather than the moment of fertilization.

and defects. It is *positive eugenics* (or, as Dr. Lederberg calls it, *euphenics,* which he defines as "the engineering of human development") that gives rise to the really vexing decisions, because positive eugenics takes upon itself the task of improving human beings—making them more intelligent, more talented, more virtuous.

Haldane, before he died, spoke enthusiastically about some of the possibilities. "One," he said, "is the deliberate provocation of mutations, probably by chemical agents, which seem more specific than X-rays and the like. This will first be attempted in tissue culture," he predicted. "And if tissue culture becomes a frequent stage in the human life cycle, it may be practicable to do it on a large scale. It may also be possible to synthesize new genes and introduce them into human chromosomes. It will still be easier to duplicate existing genes . . . There is still another possibility. No doubt, in our evolutionary past, we lost capacities which we should value, for example olfactory capacities, and the capacity for healing with little scarring which is associated with a loose skin."

Haldane suggested that we might acquire some of these capacities from animals. "Hybridization with animals," he admitted, "is probably impossible, certainly undesirable by present human standards." But other scientists, had been "able to introduce small fragments of the genome of the species of fly into another with which it gives sterile hybrids, and the same has since been done with bacteria. Such intranuclear grafting might enable our descendants to incorporate many valuable capacities of other species without losing those which are specifically human. Perhaps even 10,000 years hence this will be a wild project, but techniques progress very rapidly."

They do. And a number of scientists are willing to predict this kind of capability in less than millennia. Dr. Bentley Glass and Dr. Edward Tatum are two of the earlier practitioners of BSP who are relatively optimistic about the prospects of imparting new characteristics to new human generations through manipulation of the genetic material. There are others who raise all sorts of cautionary flags at this thought—the most outspoken, perhaps, being Dr. Commoner. These critics point out the strict limits of molecular intervention, the oversimplicity of "the DNA dogma," the complexity of the cell, the importance of all its ingredients interacting through an intricate course of development. Dr. Commoner is right enough to emphasize this complexity of interaction and its changing nature, but the nuclear material does seem to be the governing factor. In one of Bonner's experiments, for instance, the nucleic acids of pea plants, given only the protoplasmic cell material of bacteria to work with, went ahead to manufacture pea-plant proteins, not bacterial proteins. When Dr. Gurdon took the nucleus from an intestinal cell of a fully developed tadpole fetus and implanted it into an egg cell, a new, normal toad resulted. Despite the fact that much of the nuclear DNA had already been switched off chemically, communication with the body of the egg cell somehow resulted in switching them on again. The power of these switching chemicals is almost too great to predict what might be done with them when their mechanisms come under man's control. Dr. Dean E. Wooldridge, in *Mechanical Man,* cites some examples:

> The Mexican axolotl, which normally lives all its life and reproduces itself as a gilled newt in the water, can be turned into a land salamander by a single dose of thyroid. Despite the fact that the axolotl has lived its life aquatically for thousands of generations, a fraction of a milligram of thyroxin, even from a sheep or fish, will bring out the latent salamander in a couple of weeks.
>
> Indeed the induction of major changes in an entire plant or animal by the influx into the cells of gene-switching chemicals is so common that it has been observed by all of us—in the metamorphosis of a tadpole into a frog or the pupation of an insect, for example. These changes can be

induced at any time by dosing the young tadpole or larva
with suitable hormones. Familiarity may blind us to the truly
spectacular nature of such transformations. The frog differs
so much from the tadpole, and the butterfly from the larva,
as to require that each cell, in its complement of nucleic acid
molecules, carry specifications for essentially two different
species of organism, with switching arrangements to turn off
one set of controlling molecules and turn on the other upon
receipt of the proper hormonal signal. In the face of such
natural phenomena, one wonders whether the fairy-tale conversion of Cinderella's white mice into footmen was so farfetched after all!

Moreover, scientists like Dr. Glass are encouraged by strides
made in what he calls "tissue-culture genetics" similar to those
that Haldane predicted. Molecular biologists and geneticists
have seen how loose DNA can make its way into the nucleus of
a bacterium—which is also a complex and many-faceted living
cell—to change its heredity permanently, creating a new strain
of bacteria. They have seen, too—not without some alarm—
that microorganisms resistant to antibiotics are capable, quite
spontaneously, of mingling their genetic material in such a
way as to bestow upon the recipients a genetic immunity to
antibiotics which they did not previously possess. Thus, despite
the undeniable difficulties and complexities emphasized by
Commoner and others, scientists like Glass, Tatum, and Lederberg believe that it will one day become possible to modify
human genetic material in controlled ways.

Haldane especially enjoyed speculating on the genetic
changes that man might acquire from animals that would
enable him better to stand the rigors of space travel:

> The most obvious abnormalities in extraterrestrial environ-
> ments are differences in gravitation, temperature, air pres-
> sure, air composition, and radiation (including high speed
> material particles). Clearly a gibbon is better preadapted
> than a man for life in a low gravitational field, such as that
> of a space ship, an asteroid, or perhaps even the moon. A

platyrhine with a prehensile tail is even more so. Gene grafting may make it possible to incorporate such features into the human stocks. The human legs and much of the pelvis are not wanted. Men who had lost their legs by accident or mutation would be specially qualified as astronauts. If a drug is discovered with an action like that of thalidomide, but on the leg rudiments only, not the arms, it may be useful to prepare the crew of the first spaceship to the Alpha Centauri system, thus reducing not only their weight, but their food and oxygen requirements. A regressive mutation to the condition of our ancestors in the mid-pliocene, with prehensile feet, no appreciable heels, and an ape-like pelvis, would be still better. There is no immediate prospect of men encountering high gravitational fields, as they will when they reach the solid or liquid surface of Jupiter. Presumably they should be short-legged or quadrupedal. I would back an achondroplasic against a normal man on Jupiter.

The genetic material of human and animal cells have already been fused, in a crude way, in tissue culture, so even Haldane's more exotic speculations are not without some experimental foundation. Few scientists seem interested in intermingling human and animal traits, but many do believe that new genes for the improvement of man will be producible, and already existing genes changeable, in the laboratory. Dr. E. Warner Kloepfer of Tulane has expressed the opinion that DNA research "can open the door to the mass production of genes, just as the discovery of vitamins and hormones paved the way for large-scale production of those items." Rostand looks forward to a time "when each human infant could receive a standard DNA that would confer the most desirable physical and intellectual characteristics. Such children," he says, "will not be the offspring of a particular couple, but of the entire species."

The quantity of DNA necessary to transform the heredity of an entire new generation of human beings, an earthful of people numbering in the billions, could be contained in the proverbial thimble (or, as it has been more precisely calculated, in a cube measuring ⅟₂₅ inch on each side). If we had such a thimbleful, it would of course be impossible to do the job with only that quantity, considering the practical problems of distributing it and administering it all over the globe.

Many scientists warn about the hazards of any commitment to a fixed genetic pattern. Like long inbreeding, it could render the individual unable to adapt to rapidly changing circumstances—perhaps the most critical quality the new man will need in the new era. Even single-celled animals who reproduce by simple division sometimes need to combine with others in order to renew their vigor. The paramecium, for instance, after a certain number of solo divisions must occasionally conjugate with another paramecium in a sort of sexual union; then he can go back solo for a while. If paramecia are prevented from conjugating in this manner, the stock degenerates and dies out. Thus too tight control by man of man's evolution could bring that evolution to an end.

All this may be something of a game, but it is a serious game. And its goal is the power to improve human beings in whatever way we wish. Hardly anyone would insist that human beings couldn't stand some improvement. But the actual power to do so in a profuse miscellany of ways presents some sticky choices. It is one thing for a scientist like Dr. Shapley to exult in the anticipation of "a growth of social wisdom and glorious survival—toward the evolution of a kind of superman." But who is it that we will appoint to play the role of God for us? Which scientist—which statesman, artist, judge, poet, theologian, philosopher, educator—and of which nation, race, or creed—will you trust to write out the specifications, to decide, like Huxley's Predestinators, which characteristics are desirable and which not?

As Dr. Beadle points out, even deciding on the superman's superficial characteristics could be troublesome. "Given the state of the world today," he says, "the determination of his skin color alone could start a war. Should he be tall or short? Maybe he ought to be bred in a size that would fit into a space capsule better than the present model. It would be a convenience if he didn't need sleep. And since he'd have to read so much, wouldn't it be useful to lengthen his eyeballs? Then he wouldn't need glasses for close work.

"Finally, having gotten him, what if conditions on earth were to change so drastically that he turned out to be as ill-adapted as the dinosaurs?

"No. Man knows enough but is not yet wise enough to make man. And therefore our best course is to assure maximum evolutionary flexibility for future generations by maintaining a high degree of wholesome genetic diversity among men."

With so many booby-traps at every step along the way, would man not be wiser to give up the idea of controlling his own evolution? "Sooner or later," says the Canadian biologist, Dr. N. J. Berrill, "one human society or another will launch out on this adventure, whether the rest of mankind approves or not.

"If this happens, and a superior race emerges with greater general intelligence and longer lives, how will these people look upon those of us who are lagging behind? One thing is certain: they, not we, will be the heirs to the future, and they will assume control."

In a scene from Shaw's *Back to Methusaleh*, Conrad says,

> Well, some authorities hold that the human race is a failure, and that a new form of life, better adapted to high civilization, will supersede us as we have superseded the ape and the elephant.
> BURGE: "The superman, eh?"
> CONRAD: "No. Some being quite different from us."
> LUBIN: "Is that altogether desirable?"

FRANKLYN: "I fear so. However that may be, we may be quite sure of one thing. We shall not be let alone. The force behind evolution, call it what you will, is determined to solve the problem of civilization; and if it cannot do it through us, it will produce some more capable agents. Man is not God's last word: God can still create. If you cannot do His work He will produce some being who can."

BURGE: (with zealous reverence) "What do we know about Him . . . ? What does anyone know about Him?"

CONRAD: "We know this about Him with absolute certainty. The power my brother calls God proceeds by the method of Trial and Error; and if we turn out to be one of the errors, we shall go the way of the mastodon and the megatherium and all the other scrapped experiments."

But the mastodon and the megatherium did not know any biology. Man knows quite a lot—enough, hopefully, to avoid an ignominious end as a "scrapped experiment." If he can substitute human purpose for trial and error, he will have earned Sir Julian Huxley's designation as "the trustee of evolution." One way or another, the trusteeship will involve control of the genetic chemicals, DNA and RNA.

When Dr. Rostand predicted the eventual synthesis of DNA, he was talking about a truly artificial DNA, a DNA designed by man, with all the atoms arranged in the desired order to produce a certain specified kind of individual. The recent synthesis of DNA and RNA is a much lesser phenomenon. Scientists, in these instances, used existing DNA and RNA as starter substances to manufacture more of the same kind of DNA and RNA out of their constituent building blocks.

Whenever a new nucleic acid synthesis has been announced, the question has arisen: Is this *it*? Is this the creation of life in the laboratory? The answer must be heavy with contingencies: It depends on what you mean by creation, by synthesis. It depends on how you define life.

What has been done so far, though of Nobel Prize caliber from the standpoint of experimental science, does not begin to fulfill the requirements for the age-old dream (or nightmare): that is, to start from scratch with simple inorganic chemicals and, without the aid or intervention of any already existing DNA or RNA, to convert them into organic chemicals, and then into living organisms.

Will it ever be possible? A considerable body of scientists think so. Many others believe—and some hope—that this achievement will always elude man.

The late Dr. Lecomte du Noüy, one of the first biologists to understand and employ the power of mathematical techniques, spent a good many pages in his book, *Human Destiny,* arguing that the laws of probability rendered it ridiculous to imagine that any biological organization so fantastically complex as a protein molecule could have come about by chance; hence it must have been designed by providence. Some years back, when the book first came out, I remember sitting up for too many hours arguing with a college classmate about the possible origin and creation of life. No matter what fairy tale of evolutionary development anyone might dream up, he insisted, du Noüy's marshaling of the facts and the probabilities had proven beyond doubt that only God could make a tree. The dust of the earth could *not* blow into the air and make anything but a cloud of dust.

The prevalent theories of the formation of stars and solar systems made it seem not at all unlikely, at the time, that the earth was a unique planet, and that the life on its surface might exist nowhere else in the universe. This being so, the probabilities were indeed all in du Noüy's favor. But new cosmologies began to be discussed. New theories about the formation of stars and planets seemed more acceptable, theories that made it seem likely that the earth was far from unique, that planets something like it, roughly the same size and roughly the same distances from their parent suns, must exist by the hundreds of millions. Theories of how the earth itself had evolved also encouraged the idea that, at a certain period of its history, the conditions on its surface and in its atmosphere were conducive to the spontaneous beginnings of life out of inorganic materials.

In the light of the new ideas, the probabilities were now all in the other direction. If in fact the early constituents of the planet were as theorized, then it became highly probable that protein molecules *would* arise. That being the case, and there being so many hundreds of millions of planets like earth in the universe, then it followed that there might be millions of places where life exists—including intelligent life, and civilizations considerably more advanced than our own. This is what has led serious scientists and serious writers to contemplate the real possibility of one day confronting life from other worlds—and to figure out ways to make contact by radio over the light-years of distance separating us.

Back in 1953, when he was still a graduate student at the University of Chicago, Dr. Stanley L. Miller put into a test tube some of the materials (a colorless mixture of gases) which the Russian theorist Dr. Aleksandr I. Oparin believed to have been present in the primordial atmosphere. To simulate the kind of cosmic radiation that might have been around he shot electric sparks through the gases. The result was the formation of organic compounds, including the kind of building blocks that form protein molecules. Others have followed up this work—most vigorously Dr. Sidney W. Fox, head of the Institute of Molecular Evolution at the University of Miami—using a variety of techniques to simulate possible primeval conditions and producing ever more complex molecules, bacteria-like spores, and even a primitive sort of nucleic acid. Many scientists see a number of alternate ways in which life might have come to pass from nonlife during the earth's long history, and have suggested a number of alternate routes to the possible synthesis of life in the laboratory.

Suppose life *were* to be synthesized in the laboratory—undeniable life, meeting everyone's criteria. Would it mean that materialism at last reigns supreme, that spirituality is dead? Materialism versus vitalism is an old issue in philosophy. There have always been those who argued that life was "nothing but" matter, the organization and structure of which, in its complexity, gave rise to what we call thought, spirit, or soul. These were the materialists. The vitalists argued that above and beyond matter there had to be some vital force, separable from, though perhaps inhabiting and animating, matter; and that without this spirit, soul, or vital force, mere matter could never manifest itself as life, and especially not as human life.

Each new basic advance in biology seems to explain yet another former mystery of life purely in terms of electrochemistry —complex configurations of atoms and molecules. Each such advance seems to be a victory for the materialists. (As 1967 was becoming 1968, Dr. Wooldridge was putting out a book arguing that biology was reducible to physics and chemistry, while Dr. Commoner was insisting, on a televised seminar, that biology would never be reducible to physics and chemistry.) But let us assume we will arrive at a day when raw, inanimate materials have in fact been taken, step by step, through an artificial evolution to intelligent life. Would this mean an uncontested victory for the materialists?

Not at all. It would only push the argument back into the "inanimate" world. If atoms can be electrochemically joined into molecules in a consistent and inevitable way because of their inherent dynamic structure and content; if molecules, under a given set of circumstances, will form a given substance; if substances consistently form specific aggregates of substance; and if aggregates begin manifesting biological properties due to the complexity of their organization; then it could mean that the vital essence already resides in the initial atomic building blocks which we mistakenly call inanimate. A stone may look pretty dead, but the atoms in it are incredibly dynamic entities, seething and dancing with "life," their electrons excitedly jumping orbits, their nuclei highly implausible packets of compressed energy in accordance with Einstein's formulation. Thus, instead of spirit being made of nothing but matter, it can be—and indeed has already been—argued that matter is made of nothing but spirit.

As put by Dr. Loren Eiseley in *The Immense Journey:* "Rather, I would say that if 'dead' matter has reared up this curious landscape of fiddling crickets, song sparrows, and wondering men, it must be plain even to the most devoted materialist that the matter of which he speaks contains amazing, if not dreadful powers, and may not impossibly be, as Hardy has suggested, 'but one mask of many worn by the Great Face behind.' "

Suppose we accept the current scientific view that life evolved on earth from inanimate matter through a series of fortuitous cosmic events. Does this acceptance render life any less marvelous? "I take a jealous pride in my Simian ancestry," W. N. P. Barbellion once wrote. "I like to think that I was once a magnificent hairy fellow living in the trees, and that my frame has come down through geological time via sea jelly and worms and Amphioxius, Fish, Dinosaurs, and Apes. Who would exchange these for the pallid couple in the Garden of Eden?"

Many who accept this whole idea with equanimity do so in the belief that, whether you call it creation or evolution, it all came about through divine intervention. Their equanimity might be shaken if the same feats could be duplicated by the hand of man in the laboratory. Yet need it be? If the divine chemicals are mixed with the divine recipe, does it make any real difference in the end product?

A man would still be man. Or would he?

Synthetic man is obviously a long way off. And the manipulation of the genes in the kind of precise detail implied in the concept of genetic surgery is just as obviously a task of staggering complexity, perhaps several lifetimes of hard work away from us. Even so, it is not too soon to start grappling with the predicaments genetic surgery gives rise to, because exactly the same predicaments will arise with other developments we may expect to descend on us momentarily. In fact, our legal and social structures are already too antiquated to come to grips with what we now know.

Take a simple procedure like artificial insemination. When it was merely a tool in animal husbandry,* hardly anyone took it seriously as a technique for people. When people started using it, some growls of disapproval were sounded, and laws were unsuccessfully introduced to regulate it one way or the other. Today, there is still anything but universal acceptance of the impregnation of a woman with the sperm of a man who is not her husband. Accepted or not, if a couple who cannot otherwise have children want to have them this way, doctors are generally willing to comply. "Children conceived by artificial insemination," says Dr. Alan F. Guttmacher of the Planned Parenthood Foundation, "often mean more to families than children conceived in the normal manner. These children are

---

* As a measure of its current use, in the year 1960 more than seven million dairy cows were artificially inseminated in the U.S.

151

wanted—often desperately wanted." The want is in fact so desperate that many thousands of babies are born in this fashion every year in the United States alone. It is estimated that up to 150,000 Americans now alive would never have been born without the use of artificial insemination. Yet, despite the commonness of the practice, its legal status is still so questionable that often the birth certificate gives no indication that the child is not the natural offspring of the mother's husband. And no wonder.

Dr. Glanville Williams, the well-known British law professor, cites a case where the Supreme Court of Ontario, Canada, ruled that artificial insemination—on this occasion, without the husband's consent—constituted adultery. Adultery was defined as "the voluntary surrender to another person of the reproductive powers or faculties of the guilty person." The judge in the case went even farther: "If it was necessary to do so, I would hold that [artificial insemination by donor] in itself was 'sexual intercourse'." A similar opinion was expressed at a scientific conference in London by Dr. Colin G. Clark of Oxford, who said, "Coming back to the question of morals: artificial insemination by a donor other than the husband has all the malice of adultery—and I think that anyone who understands the moral meaning of the word adultery is bound to reach that conclusion."

More recently an Illinois judge declared the practice adultery whether the husband consented or not—and, moreover, ruled that the child in such a case was illegitimate. Does this, Williams wonders, brand the anonymous donor of the sperm an adulterer? And could the husband name him as correspondent in the divorce proceedings? For that matter, could the donor's wife divorce *him* on grounds of adultery? True, the donor is always assured of anonymity, but could a court force the doctor to reveal his name? On the other hand, since it is the doctor who directly administers the semen, thus personally and deliberately impregnating the lady, should it not be the doctor who is named the adulterer? "This solution may look a little odd," says Williams, "if the physician is a woman."

Artificial insemination is neither forbidden by the law nor condoned by it. Except for the statute just passed in 1967 legalizing it in the state of Oklahoma,* the law has simply failed to deal with it at all. The public, including the legal and juridical public, has had to muddle through as best it could. As biological advances become more radical in nature, muddling through simply will not do. The next step in this direction, for example, might be sperm banks. And every bank will generate a briefcase full of legal headaches.

In artificial insemination as practiced so far, the couple has nothing to say about the selection of the donor. The doctor ordinarily uses his own discretion. He selects a young, healthy man, as often as not a medical student, and one who is roughly the same physical type as the husband. He picks a man who, as far as can be ascertained, is free of congenital defects—and preferably a married man who has had at least one normal child of his own. The donor has no idea who is to be inseminated, nor does the couple know the identity of the donor. "Among the reasons for this secrecy," says Glanville Williams, "are the desire to protect the donor's reputation (think of the

---

* During the lengthy debate that preceded passage of the statute, one legislator opposing it said, "I don't think people should be treated as cattle. This is putting the human being on the same level as four-legged animals."

Another said, "It might be God's will that a couple not be able to bear a child."

A third thought the bill sounded as if it "sort of sanctioned adultery," while another warned that "Oklahoma could become known as the Sweden of the United States."

repercussions for his family if his adventures in paternity became common gossip!), and to eliminate the risk of the wife transferring her affection to the donor."*

The current procedure is deplored by some. To the late Dr. Hermann Muller it seemed a tragically haphazard way of doing things. He proposed in its place "germinal choice," a plan which a number of other scientists, including Sir Julian Huxley (who calls the idea "pre-adoption"), find merit in. A truly enlightened couple, in Muller's view, would forego the purely egotistical satisfaction of imparting their own hereditary traits to their children, and instead choose, when the choice becomes available, from the sperm of "those whose lives had given evidence of outstanding gifts of mind, merits of disposition and character, or physical fitness" in order to endow their children with "the kind of hereditary constitution that came nearest their own ideals."

Muller's suggestions that such sperm banks be established is rendered quite practicable by the development in England and France of simple and reliable techniques for putting sperm in the deepfreeze for indefinite periods of time without apparent damage or deterioration.** At the State University of Iowa some years back, Dr. Ralph O. Bunge and Dr. Jerome K. Sherman conducted an extensive series of pioneering frozen-sperm ex-

---

* In a novel called *Test Tube Father,* author Francis Silvin explores the emotional problems of a couple who decide to have a baby via the artificially-inseminated sperm of an anonymous donor. The father suffers all kinds of jealousy and the mother has obsessive dreams about the child's real father.

** The big breakthrough came via serendipity. Dr. Parkes, much honored for the work he and his colleagues did at London's National Institute for Medical Research, later told how the key discovery had come about because some egg-albumen and glycerol accidentally got mixed up with the test solutions. Glycerol turned out to be the key to successful freezing—without destroying or damaging—the sperm.

periments on animals. Satisfied that the method was safe and effective, they began using it on human mothers. Others elsewhere have since followed suit. Many healthy children are alive today as proof of the validity of this new technique. There was even a report out of France to the effect that a lady biologist had borne two children years after her husband's death in a car crash, by having herself artificially inseminated with his cold-storage sperm. The report turned out to be erroneous. But it was a perfectly credible story. A dairy-farmer friend of mine tells me that some of his prize cows were born from bull semen frozen more than ten years ago. At the University of Michigan, Dr. S. J. Behrman and his associates rendered twenty-nine women pregnant by sperm kept frozen for two and a half years. "What we have done so far," he says, "is to freeze the male cell through which life has been passed on over the centuries. There is every reason to believe that this suspension can be prolonged indefinitely."

The immediate promise of frozen sperm will be—rather than the great public storehouses envisioned by Dr. Muller—private sperm banks. Some men are afflicted with oligospermia, a sperm count so low that there are never enough viable sperm at any given time to render the man fertile. By freezing and storing his sperm, however, he could save up enough to become the father of his own children after all. A man might want to bank his sperm for other reasons, too. He might be facing an operation that would render him sterile—an operation for, say, cancer of the prostate. With a good supply of viable sperm in storage, he could not really be rendered infertile. Or suppose a man were going off for an extended period of time—perhaps to war, or on a journey to one of the planets. In his absence his wife, with the aid of the family doctor, could bear a whole family of his own children. (A very jealous husband might then worry that he had provided his wife with a perfect opportunity for infidelity. Though he left behind his sperm, would he still be plagued by doubts that the children were really his? Such are the plots for the soap operas of the future.)

Muller deemed this attachment to having one's own children in the traditional sense a conceit which a forward-looking man ought to discard in favor of the greater satisfaction of knowing that a child "was a product of his deliberate volition, rather than of his reflexes" and "embodies the best genetic, as well as the best environmental influence that he was able to provide it with."

Others besides Muller have been pushing for sperm banks. Physicist Ralph E. Lapp, for instance, has strongly urged that pools of sperm be collected, "refrigerated and kept in lead-covered Fort Knoxes and dispersed throughout the nation" to insure a supply of undamaged sperm in the event of nuclear war. It was Muller's belief that sperm banks and germinal choice would become a necessity if only because the very complexity of the problems science is creating will require an improved brand of human being to deal with them. "Our Stone Age constitutions," he declared, at an Ohio Wesleyan symposium, "are being sorely stretched in trying to adapt to the unprecedented complications of civilization and of the world as seen by modern science, to the need to feel brotherhood for three billion people, and to the responsibility of guiding without disaster the use of the enormous powers that scientific technology has created. . . . The need is both for better brains in depth and breadth, with all the faculties accessory thereto, and for warmer hearts, which allow men to find more genuine fulfillment in actions that serve humanity at large."

The public sperm banks urged by Muller and Lapp might be operated municipally, or they could become nationwide or even worldwide to give the parents the broadest possible choice of donors. I don't know who would be charged with running all this. Perhaps, as Campanella suggested in his seventeenth-century Utopia, *City of the Sun,* each government would have to set up a Ministry of Love to watch over marriage and procreation.

Muller's specific plan calls for a full dossier on each sperm donor, including their relatives and the kind of lives they led. This is a most radical departure from the donor anonymity hitherto insisted upon. Such a departure would certainly exaggerate all the legal and emotional risks involved. If the donor were known, might he really be sued, in the absence of further legal clarification, for child support? Or have the terms of his will contested by descendants he never meant to be his heirs? In case of careless bookkeeping or registry, might there be the possibility of someone unknowingly marrying a near relative? Or a danger of vicarious incest?

If a young man or woman knew the "real" father, might he or she be tempted to set up a true filial relationship with him rather than with the social father? Or might the mother begin to feel more attached to the genetic father than her husband might like? She did, after all, admire his qualities sufficiently to want her children patterned after him. On the other hand, a woman—if she could select whatever genetic endowment she desired for her children—might then feel freer than ever to marry whomever she happened to fall in love with, without having to think about genetics at all.

Earlier I referred to the character in *Baby Want a Kiss* who proposed the peddling of Celebrity Seed. If germinal choice were to become a routine practice, celebrities might indeed be tempted to cash in on their seed. When the time comes, it might seem no more reprehensibly mercenary to do this than to endorse a hair tonic or cigarette. Medical students even today sell their sperm for somewhere between $10 and $25. Why shouldn't a celebrity charge more? Such a practice might totally frustrate Muller's plan, of course, since a fad for a given movie star or baseball player or rock singer, or for any hero of the moment, would not guarantee any improvement of the race.

157

To minimize the danger of troublesome emotional entangle-ments, or of transistory personality fads, Muller—who perhaps remembered with some embarrassment that he once thought of Lenin as a greatly-to-be-desired contributor of sperm—sug-gested that, ultimately, when sperm banks are operating on a large scale, germinal choice should be made only from the sperm of donors no longer living and which have been stored for at least twenty years. To help couples make their choices more wisely, he would also provide professional counseling. Thus, across time and space, guided by experts, a mother could select as father of her child a Hindu philosopher, a Scandi-navian athlete, a Congolese poet, a French astronaut. Some biologists, among them Dr. Medawar and Dr. Berrill, warn that this kind of genetic selection is unpredictable, and that the children will not necessarily be like the fathers. "Artists cannot be counted on to breed artists," says Berrill, "nor do astrono-mers breed astronomers. Nor can the inheritance of general intelligence be predicted. Unusually intelligent parents can produce human vegetables as readily as do other couples, while individuals of exceptional merit tend to crop up everywhere in ordinary run-of-the-mill families." But Muller insists that the *chances* of getting worthwhile children are better if you select a worthwhile parent, and moreover, that germinal choice will result in the long-range improvement of the human race.

Some geneticists dispute even this on the grounds of a para-doxical phenomenon called heterosis. Each of us gets each of his hereditary traits not through a single gene but via a pair of genes. Most hereditary diseases are recessive—that is, we get the disease only if *both* genes in the pair are defective. If only one of the pair is defective, it actually appears to confer advantages on the inheritor in the form of "heterotic" vigor. The best-known example of heterosis is the case of sickle-cell anemia. A pair of sickle-cell genes spells disease. But a single defective gene does not; in fact, it provides some protection against malaria.

Because of heterosis, then, it would appear that eliminating all the "bad" genes might also eliminate many of the sources of human vigor—the ironic result being people who are less fit rather than more fit, and a race deteriorating rather than improving. But heterosis, at this stage, is more a theoretical concept than a proven phenomenon. Even when proven, it will not really negate Muller's proposals—which are not drastically different from what Plato proposed in *The Republic*—i.e., to "learn from the breeders of hunting dogs and birds of prey . . . They single out from the herd the ones that excel the others."

There is little doubt that cattle breeders in our own day, for example, have been able to create considerably improved livestock through the use of artificial insemination (and have virtually banished the sexual experience from a goodly portion of the world's bovine population in the process). The breeders of cattle have been able to do this because they can easily agree on what constitutes "improvement"—better beef, higher milk production, and the like. If the breeders of human beings could agree on what was desirable, there is no reason why the more desirable traits could not be attained to an increased degree in the same rough-and-ready fashion.

Genetic choice becomes considerably enlarged, of course, when Dr. Hafez's egg-implantation techniques begin to be applied to human females. As a matter of fact, Dr. James L. Burks at the University of Chicago has succeeded in freezing, then unfreezing, human ova, and fertilizing them *in vitro*. With artificial inovulation as available as artificial insemination, with egg banks as well as sperm banks to draw from, either a genetic mother or a genetic father, or both, could be selected for the prospective child. Then both men and women could ignore heredity in their choice of mates.

Women who were willing to be superovulated by hormones, as Hafez's cows are, could provide eggs for many other women. A barren woman who could never hope otherwise to

159

be pregnant could be afforded the fulfilling emotional satisfaction of having babies after all. A woman with serious heart trouble whose doctor feared to let her bear children might, by donating an egg, have her child carried by someone else. If inovulation were to become an accepted practice, it might also be acceptable for women to hire out as surrogate mothers just as they have traditionally hired out as wet-nurses.

For whatever purposes inovulation were performed, custody cases would become even thornier. "If a woman bore a child that was not genetically hers," asks Jean Rostand, "who would be the actual mother? Would it be she who carried the child or she who furnished the germ cell?" Glanville Williams carries the conundrum a step further. Suppose Mother A were artificially inseminated with frozen sperm from a donor no longer living, then the fertilized egg were transplanted into Mother B, who actually gave birth to the child. Puzzle out the parenthood of *that* one.

In animals, not only eggs but entire ovaries have been successfully transplanted. The resulting offspring is genetically like the mother from whom the ovary was taken rather than the mother who carried out the pregnancy. In some dog experiments, defective and atrophied ovaries taken from elderly females seemed to rejuvenate spectacularly upon transfer to healthy young females. No one is certain this operation would work on people—though theoretically an ovary, consisting of embryonic tissue, would be less likely to be rejected by the body than other types of foreign transplants.* But if it did work, it would add to the possibilities, and to the problems.

There is no telling, then, what variety of choice the girls and boys who are little boys and girls today might conceivably have

---

* According to the British legal publication, *The Solicitors Journal,* a well-known British gynecologist was ready to perform this operation on a woman, but, on being informed by the Medical Defense Union that any resulting offspring would be considered illegitimate, he decided not to proceed.

open to them by the time they are ready to become mothers and fathers. They could of course continue as of old, assuming both partners were fertile. They could have the wife artificially inseminated with donor sperm. They could have her artificially inovulated, then inseminated with the husband's own sperm in the conventional manner. Or they could have her first artificially inovulated, then artificially inseminated. They could buy an already fertilized egg, if Dr. Hafez's prediction were to come true. They could use their own egg and sperm and have the baby raised partially or all the way *in vitro,* depending on the state of the art. Or use some other egg and sperm and still have the baby raised *in vitro.*

So it is obvious that we can stop far short of the realms of fantasy where people are raised in tissue culture—and, stopping far short of them, still possess powers for the relatively near future which are fraught with hope and peril. In our own time, uncertainty and insecurity already plague most forms of human relationships. Now we are entering an era where children may be born of geographically separated or even long-dead parents, where virgin births may become relatively common, where women will give birth to other women's children, where romance and genetics may be separated, where some few favored (or ill-favored) men may father thousands of babies, where a permit may be required in order to have a baby.* Can the traditional family—already a shaky institution—survive in

---

* Dr. Francis H. C. Crick of Cambridge, in a discussion on eugenics and genetics, asked the question: "Do people have the right to have children at all? It would not be very difficult for a government to put something in our food so that nobody could have children. Then possibly—and this is hypothetical—they could provide another chemical that would reverse the effect of the first, and only people licensed to bear children would be given this second chemical. This isn't so wild that we need not discuss it. Is the general feeling that people do have the right to have children? . . . I think if we can get across to people the idea that their children are not entirely their own business and that it is not a private matter, it would be an enormous step forward."

the midst of all this? Do we want it to survive? If so, how will we insure its survival? If not, what will we substitute for it?

Not only the foundations of family life, but many of our laws, the themes of art and literature, and much of our ethics, morals, and even politics, are based on premises which have been taken for granted from time immemorial. Soon these premises are likely to be challenged.

The nature of human relationships must be thoroughly re-examined—and, some think, reconstructed—if we are to manage sensibly the new controls that scientists will hand us as a result of their exploration of prenativity.

We are well aware that, even before the advent of these controls, contemporary human relationships are fragile and tenuous, so much so that many behavioral scientists have been talking with some enthusiasm about the desirability of "arranging human relationships." This talk grows out of a realization, as more and more work is done with groups of normal people—just plain folks who are neither under psychiatric care nor think they need it—that most human relationships are, in the words of Dr. Farson, "appallingly superficial." Individuals locked inside themselves have a desperately hard time communicating with others—even with their closest friends and associates, even with (or perhaps *especially* with) members of their families. They talk, of course. They "present themselves" to one another. They play roles of various sorts. But they almost never reach one another in any deep, essential way on a level of unguarded human intimacy. Most people consciously try to reach others in this fashion, but in their fumbling explorations, fail to find their way through the interpersonal barrier. Many fear to make themselves vulnerable by exposing their true feelings—sometimes even to themselves. Yet psychiatrists believe that everyone hungers, whether he knows it or not, for such "basic encounters"—for without them he can never achieve his full humanity. "Millions of Americans," says Farson, "have never had, and never will have, in their entire lives, one moment of intimacy with another human being— one moment in which they could be honestly, authentically, genuinely themselves . . . We need new designs for living which will make emotional intimacy possible, which will encourage it—not force it, but encourage it . . . It will be our job as professionals to arrange situations or conditions in which people can helpfully encounter each other—in which they may enjoy deeply personalizing experiences."

One of the popular experimental techniques is the bringing-together of small groups of people, usually people who have never before met one another, for a long weekend of talk in an arranged, artificial situation. A network of such "encounter groups," variously named and organized, now exist—some of them one-shot events, others continuing on a longer-range basis. Some are sponsored by nonprofit research organizations like the Western Behavioral Sciences Institute in La Jolla, California; some by commercial concerns like the Human Development Institute of Atlanta, recently acquired by Bell & Howell. Many of the groups are kept in touch through the auspices of the National Training Laboratories' Institute for Applied Behavioral Science in Washington. People in encounter groups tend to be diffident at first, but they soon find themselves loosening up and talking about themselves and one another with a kind of honesty and intimacy that surprises them. They wind up feeling very close to one another, often each person confessing that, for the first time in his life, he has reached and been reached by other human beings in a truly basic encounter. Almost everyone, at one time or another, has found himself talking more freely and confidentially with, say, someone who simply happens to be sitting next to him on a plane, than he is ever willing or able to do with a working colleague or a next-door neighbor who is seen almost every day. Yet even those familiar with this phenomenon are amazed to learn how far it can go in an encounter group. For people who discover they can achieve this kind of communication with perfect strangers on short notice, it seems doubly tragic that they are unable to attain it in their home situations with those they love—or once did.

"That a man and a woman who are attracted to one another should agree to share life in mutual devotion," the late anthropologist Dr. Robert Briffault once said, "is one of the most satisfactory arrangements which social culture has brought into being. But it is an arrangement the success of which is not provided for by natural instincts." He added: "The chief condition for the success of marriage is to know what one is about."

Continued failure to communicate builds up a massive cumulus of frustration, resentment, and unhappiness, especially when the noncommunicants happen to be married to one another. That is one reason why the divorce rate is so high, and would be even higher if a lot of people did not work hard to "make a go of it." Unfortunately, the go they make of it frequently amounts to nothing more than a borderline accommodation to a minimally tolerable arrangement where both parties feel trapped and isolated at the same time. On every side we see husbands and wives, unable to get through to one another, struggling to get through to their children—who have, meanwhile, grown increasingly unreachable as values of all kinds become more nebulous and more difficult to justify by an appeal either to tradition or to reason.

The failure of communication, along with a concomitant failure of adult values, is certainly one major reason for the much-publicized "generation gap," which is believed to have grown wider than ever in recent years. It is hard enough for confident parents, even in a more stable era, to impart what is traditionally assumed to be their superior, experience-based wisdom to growing children at a stage when they are just beginning to feel a spurious adulthood, when they mistakenly assume that, because their muscles and sex organs and voice-boxes have taken on adult dimensions, so have their values and judgment. But in a chaotic time like ours, how persuade teenagers to "behave"—whatever that is—or even that they ought to? Many in the motorcycle and surfing crowds openly admit that they do some of the outlandish things they do—such as wearing iron crosses and Nazi swastikas—mainly because they know it upsets their parents. The moral relativity that afflicts our shifting society further confounds every confusion, aggravates each insecurity, and renders communication even more difficult—especially when the matters to be communicated have themselves become so ambiguous. Paul Goodman suggests that "perhaps there has *not* been a failure of communication. Perhaps the social message has been communicated clearly to the young man and is unacceptable."

One set of confusions that assails both generations almost equally is the fluid situation that characterizes our attitudes toward sex. "If we are to provide the impetus for surmounting the trials and obstacles of this most difficult period in history," says Dr. Margaret Mead in *Male and Female,* "man must be sustained by a vision of the future so rewarding that no sacrifice is too great to continue on the journey towards it. In that picture of the future, the degree to which men and women can feel at home with their own bodies, and at home in their relationships with their own sex and with the opposite sex, is extremely important."

One of the most powerfully disruptive series of changes that comes about during adolescence is the rapid development of sexual characteristics—the whole spectrum of psychophysiological happenings that make a boy aware that he is becoming a man and a girl that she is becoming a woman. The new awareness of the sex organs and other bodily changes is accompanied by the first strong and often baffling stirrings of sexual desire. Though today's children are surrounded by, and soaked and steeped in, sex of a sort, there is a real dearth of information and guidance they can really count on. "They are trapped," says Paul Goodman, "by inconsistent rules, suffer because of excessive stimulation and inadequate discharge, and become preoccupied with sexual thoughts as if these were the whole of life." They need help, they know they need it, and they want it. They hunger, says Dr. Mary S. Calderone, executive director of the Sex Information and Education Council of the United States (SEICUS), "for straight and open *talk,* for straight and open *information,* and most especially for straight and open *feelings* on the part of adults." Unfortunately, parents seem incapable of this kind of straight and open communication, even more in this area than in others. They are ill equipped to be of real service because they are afflicted by so many of the same confusions. Events have moved too fast for them.

At this point in time, long before the more bizarre eventualities we've discussed have come to pass, a quick look around is all it takes to confirm that a startling transformation has taken place in our attitudes toward sex. It is hard to say what is "normal" and what is not any more. All sorts of behavior which only a few years ago was considered wrong, or at least questionable, now seems acceptable to many people. Sex is talked about much more openly, and there is no reason to doubt that it is practiced much more freely too. Contraception is becoming easier and more available, and medical advances have made venereal disease a less fearful specter than it once was—though the rate still seems to be going up among young people. In the scientific laboratory, sexual activity is studied clinically, recorded and measured by instruments, and photographed in color by motion-picture camera; and, in this context, it is perfectly legitimate for men and women of various ages, alone or with partners, with or without the aid of artificial devices, to perform sexually and even earn a modest fee for their contribution to scientific knowledge.

Playwrights and novelists feel freer than ever to describe any kind of sexuality they can imagine in whatever terms seem suitable to them. Books once available only by mail in plain brown wrappers now flourish on paperback racks in Woolworth's and your local pharmacy. Doesn't it seem a long time ago that the obscenity of that innocent book, *Lady Chatterley's Lover,* was argued in the courts? Now there sits *Fanny Hill,* in any drugstore, at any airport, surrounded by, say, *The Story of O* and *Naked Lunch* on one side, and, on the other, by Krafft-Ebing without the Latin, and the unexpurgated Marquis de Sade. Sex in the movies leaves little to the imagination. Some films take place entirely in the bedroom, and the sex may be marital, premarital, or extramarital, but in any case there is seldom so much as a curt nod toward the old proprieties. Now that the underground film seems to be surfacing, the latitude becomes even wider. Ads in respectable family magazines are full of women in various stages of undress. Dimly lit bunny clubs flourish across the country, and even at high noon, if you happen to be in the right locale—especially in California—your luncheon order may be taken by a waitress stripped to the waist. Some of the more daring evening dresses show more flesh than bathing suits used to. Homosexual organizations are expanding.

On the college campus, where a lingering goodnight kiss at the dormitory door was once considered a bit wicked, premarital sex—while not a universal pastime—has become a taken-for-granted activity.* Around a few campuses, free-sex

---

* It is hard to remember that, as recently as 1960, the University of Illinois fired a biology professor for suggesting that premarital sex might be ethically justifiable.

clubs featuring nude parties have sprung up, some of them institutionalized to the point of electing officers and issuing membership cards. (Similar happenings have been reported in Japan and Austria.) "Erotic stimulation is all around us and pouring in on young people from every imaginable source," says Dr. Calderone; and Dr. Lester A. Kirkendall, a professor of preventive medicine, emphasizes the new "free-choice situation" into which adolescents find themselves increasingly thrust. Even Henry Miller, that celebrated celebrant of sex, has been grumbling aloud that things have gone too far.

Everything's *à Go-Go,* and Everyman's a swinger—or seems to want to be. And Everywoman, too—with the very young apparently in the vanguard. There is no way to document for certain the degree of correlation between public displays and private acts—to what extent, for instance, a teenie-bopper's gyrating to rock music is merely an innocent letting-out of excess youthful steam. But, if we can believe Frank Zappa— who, as leader-composer of one of the nation's top rock groups, the Mothers of Invention, ought to know—"the level of involvement with today's music is quite amazing. One example: Groupies. These girls, who devote their lives to pop music, feel they owe something personal to it, so they make the ultimate human sacrifice. They offer their bodies to the music or its nearest personal representative, the pop musician. These girls are everywhere. It is one of the most amazingly beautiful products of the sexual revolution." Nor does the shedding of sexual inhibitions appear to be restricted to "groupies" alone. At a well-publicized public panel in New York, a teen-age girl told an audience of hundreds—most of them professionals exploring teen-age sex attitudes: "You can't tell a sixteen or seventeen-year-old she can't have intercourse. She's going to do it anyway." Though one assumes that this girl's viewpoint is not typical of American teen-agers, a surprising number of youngsters do seem willing to make equally unabashed remarks in public. A Yale University study of sex among teen-agers disclosed that, during the year the survey was made, one out of every six girls in the state of Connecticut between the age of thirteen and nineteen was illegitimately pregnant.* There were no figures for the years below thirteen. But the manufacturers of rock-'n'-roll type "love ballad" records and other Go-Go items admit they are already striking it rich by deliberately aiming at the four-to-twelve-year-old market. After the teenie-bopper, the micro-bopper?

Even before this sexual revolution, there were problems aplenty in interpersonal relationships. A chronic failure of communication is bound to create a nagging sense of discontent and insecurity, under the best of circumstances. Add an ingredient—the prevailing liberalized attitudes toward sex—and you add to all existing confusions and insecurities. Dependable standards of fidelity are getting harder to come by. How are married couples to fix them, even for themselves, with convincing validity, let alone set standards of judgment to apply to other people? And in their own state of ambiguity, what standards do they fix for their growing children—and how do they make them credible?

In a society where we live with such transitional sexual values, where grandparents usually live elsewhere so that children scarcely know them, where a high degree of geographical

---

* The 1-in-6 figure—a projection of what was considered a statistical sample of the population—when originally put forth by the Connecticut State Department of Health in 1966, was greeted by incredulity. But authorities pointed out that this was not inconsistent with figures from many other states, and some statisticians believed the 1-in-6 figure was too low because (1) it probably underestimated the number of girls who lose their babies through abortion; (2) it did not include unmarried mothers who assume a married name and invent a mythical father; and (3) it omitted girls who have their babies in other states though they become pregnant in Connecticut.

mobility exists so that daddies travel a lot and families move often, it takes no sage or savant to point out to us that at this moment in American social history, the institutions of marriage and the family are in serious trouble.

I would certainly not want to see us beat a full-scale retreat to the old prison of puritanism and strict censorship, where sex becomes something furtive and dirty. Nor do I disapprove of the sex research which has finally begun to be carried out—by the Kinsey group, by Masters and Johnson, and by a number of others. It is about time we got some reliable information about sex to replace the mythology. Dr. John Gagnon and Dr. William Simon, two University of Indiana sociologists, decry the fact that so many people who are enthusiastic about sex education for the young have no enthusiasm for the kind of research that will provide us the knowledge to *teach* the young. This state of moral confusion puts additional stresses on the family, accelerating the erosion that is already in process.

"What is the present situation as regards marriage and the family?" asked that wise anthropologist, Dr. Bronislaw Malinowski, on a BBC broadcast back in 1931. "Traditional morals and the legal framework of domestic life are undergoing disquieting changes. There is a crisis in marriage and there is a great deal of noise about it. Let me add at once for your comfort that there is more of the noise than of the crisis."

It is easy enough for us to smile and comfort ourselves. "See," we can tell ourselves, "this happens in every generation. What Dr. Malinowski said in 1931 probably applies as well today. Did he himself not poke fun at the predictions of Professor J. B. Watson and his Behaviorists, who seemed happy to assure everyone that marriage was on its way out?"

Agreeing with Malinowski twenty-five years later was Dr. Ashley Montagu who declared firmly: "The biologically normal state of the basic human biologic group is the family, consisting of parents and their children, and we can be quite certain that

170

nothing will ever permanently change that fact. Since the family is based upon marriage it is similarly possible to predict the permanence of that institution."

But sexual attitudes had not undergone nearly the same kind of wrenching metamorphosis as they have in the intervening years. And the current moral flux has taken place entirely in the context of a continuing trust in the relative immutability of what we so fondly and familiarly call the Facts of Life. When the new and entirely mutable facts of life take over, what even greater flux will ensue?

The moral sanctions of religion once served as a sufficient guide for most people. But for more and more formerly secure believers, the message has not gotten through, and even theologians and clergymen have been expressing massive doubts about the grounds on which these sanctions have been based, and are, with great outpourings of words and works seeking new grounds for old tenets. And there seems to be evidence for an apparent willingness to drop the old tenets if the grounds they find are unconvincing.

In 1966 the British Council of Churches appointed a committee, headed by the Rev. Kenneth Greet, to compile a report, with recommendations, on the current state of sex and morality. When the committee turned in its report, a lot of people were shocked. Considerably loosening the standards of church tradition, the committee made it clear that it did not necessarily condemn all cases of either premarital intercourse or adultery. Moreover, it called for birth-control advice for unmarried teenagers, and liberalized attitudes toward abortion and homosexuality. Ardis Whitman, in *Redbook,* quotes a young chaplain's advice to his students at an American women's college: "Sex is fun, and there are absolutely no laws attached to sex." And an Episcopal theologian: "If people do not believe it is wrong to have sex relations outside of marriage, it isn't." And Dr. Gibson Winter, professor of ethics and society at the University of Chicago's Divinity School: "Personal responsibility is going to be the essence of any morality of sex in our time." This is not to say that this new viewpoint has replaced the old religious one—to which most church people still adhere—but it is sufficiently widespread to make many laymen feel that, sexually, they are on their own. All this may be good and healthy, but for the ordinary man or woman looking for guidance, if the fear of God is not enough, what then?

We have already mentioned some secular fears which have always existed, even among the nonreligious, to act as deterrents to infidelity and promiscuity: fear of venereal disease; fear of causing pregnancy or of getting pregnant; fear of being found out and having to face social disapproval. Another deterrent has been mere inconvenience—the nuisance of seeking a private place, the annoyance and mess of contraceptive precautions.

We all know that, in point of fact, neither the fears nor the inconveniences ever served as truly effective deterrents for many people. Suppose, then, we reach a time—and we may reach it soon—where venereal disease is no longer any threat at all, where contraceptive methods are so cheap and convenient as to remove any fear of unwanted pregnancy. With physiological immunity thus assured, we can also suppose that, with changing attitudes, there may also be social immunity; that is, even if one is found out, no one will care. In fact, there would be no point in secrecy at all. The very wide-openness of sex might of course rob it of much of the erotic excitement that now surrounds it—with the risk that the resulting blandness of the experience might render it a bore.

"However closely human sexuality is tied to biological processes," say Drs. Gagnon and Simon, "it is also subject to the molding impress of the sociocultural to a degree surpassed by few other forms of human behavior." They contend that "all human sexual behavior is scripted behavior"—that is, our sexual attitudes are instilled by society rather than inborn. One remembers the easygoing sex of *Brave New World*. In a more recent novel, *The Harrad Experiment,* author Robert H. Rimmer portrays quite convincingly the ease with which a group of college students cast off their old morality to take freely to sex in an environment where nobody thinks it is wrong. Older people might find themselves similarly adaptable; wife-swapping clubs have existed for some time. And Dr. Albert Ellis, head of New York's Institute for Rational Living, has suggested that even for happily married, nonpromiscuous types—if they were mature enough and secure enough to handle it—an occasional extramarital affair might be an enriching experience. With changing circumstances we could all find ourselves living in a quite different moral milieu.

For any man or woman living in this changed moral environment, the opportunities for sexual adventures will obviously increase—though enhancing the opportunities may diminish the adventure. For any husband or wife so inclined, the temptation to philander may be overpowering. The man or woman who is not personally tempted, but who is subject to jealous apprehensions, is bound to become more uneasy with the awareness that the party of the second part may not be resisting temptation with equal success. A jealous man or woman traditionally has at least had the sympathy of friends. But they might find that most of their friends think it is absurd to expect anyone to be faithful. The effect of all these pressures would vary with the individual, of course, but in the case of a marriage already precarious these added concerns could easily finish it.

With old fears replaced by new freedoms, do the foundations of fidelity then fall? Does fidelity become an outmoded concept? And if sex outside the marriage bed is O.K., what happens to marriage itself? Do we marry for love, companionship, security? And are these lasting? Should we be prepared to change partners whenever there is a feeling on the part of either one that it's time for a change? Are the legal bonds of marriage nonsense? "Marriages," says Henry Miller in *Esquire*, "are made in heaven, and they are unmade in heaven, and then they are over! They last only as long as there is beauty in them, a validity and a reality in them."

Is the ideal, then, to be a purely personal arrangement without law or ceremony, a companionate arrangement like the one between Jean-Paul Sartre and Simone de Beauvoir, for example? Even Dr. Mead points up the enormous hazards in the way of staying married for life, especially in the United States, where marriages undergoes extraordinary strains because of the great romantic expectations it must uphold. "The ideal is so high," says Dr. Mead, "and the difficulties so many, that it is definitely an area of American life in which a very rigorous reexamination of the relationship between ideals and practice is called for."

But what about children? How are they to be reared? How are they to be given any sense of stability if we seriously modify the nature of marriage? Henry Miller does not ignore that problem. It worries him. But he suggests that there is plenty of damage being done to children under present circumstances too. How can children expect to get any love, he wants to know, from parents "who don't show any love, even to each other, or to mankind in general? Whether the family stays together or not, the children are still suffering." In his view, we can't trust children to parents anyway, and ultimately everyone's children are society's concern. So the maintenance of marriage merely as a framework for child rearing is also of questionable validity.*

If the sole reason for marriage as an institution were to be based on this premise, then marriage would indeed soon be-

---

* Henry Miller's credentials as an advisor on marriage and the family would hardly be universally acceptable. Nevertheless, similar sentiments have been expressed by sociologists and educators.

come *passé*. Children have always been thought of as products of the marriage bed. "Moved by the force of love," Père Teilhard de Chardin once wrote, "fragments of the world seek out one another so that a world may be." The fragments of the world he was talking about were the sperm and the egg—the sperm fresh-sprung from the father's loins, the egg snug in its warm, secret place; the propelling force being conjugal love; the new world being the child itself.

But the force of love may henceforth have little to do with it all. The crucial fragments of the world may simply be taken out of cold storage. Even if the scientist or technician who brings the fragments together managed to maintain an attitude of reverence toward the life he was thus creating, love in the old sense would have vanished from the procreative scene.

Assuming that the father's own sperm and the mother's own egg were used, the mere fact of conception outside themselves, conception in which they did not personally participate, might make a vast difference in their later attitudes toward their children. If the sperm and/or egg belonged originally to someone else, it would add to the impersonality of the whole transaction. How much of any mother's feeling toward a son is bound up in the physical fact of having carried him inside the womb for those long intimate months, nourishing him with her own body, drawing emotional sustenance from his presence there, fulfilling herself physiologically and spiritually as a woman? With this gone, would her maternal feelings be the same?

Every psychiatrist knows that parental feelings toward children often leave much to be desired.* One sign of this in our time is the recent addition to the medical lexicon of an ugly

---

* Says that once unhappy child, Jean-Paul Sartre: "There is no good father, that's the rule. Don't lay the blame on men, but on the bond of paternity, which is rotten. To beget children, nothing better; to *have* them, what iniquity!"

new term, "the battered-child syndrome." In case after case, battered children are brought in to doctors or hospitals. The batterers are their parents—one or the other of them—parents who usually do not consider themselves disturbed or in need of psychiatric help. Sometimes the pretext is that the child has fallen and hurt himself. Sometimes there is no pretext at all. The mother may have beaten the child because he refused to eat, or simply because she couldn't stand him—and beaten him with fury and without mercy. Big brawny fathers have clouted infant daughters around with the full force of their fists. Children covered with bruises and welts, children with multiple fractures, children with external and internal bleeding—all battered-child victims. This tragedy has become so common that doctors have begun to look for it and report it to authorities, and committees have been appointed to study it. Only a small percentage of the actual cases are ever seen or reported. "If we had the real figures," says Dr. Frederic N. Silverman, a Cincinnati radiologist on one of the committees, "the total could easily surpass auto accidents as a killer and maimer of children." And reports out of Europe indicate that the phenomenon is not indigenous to the United States.

There are of course always those people who very much *want* children. They not only love their own, but are capable of giving genuine love and affection to children born of artificial insemination, even if the sperm is from an anonymous donor. They can love a child who has been totally adopted, not in any sense their genetic product, not carried by the mother nor sired by the father. Might the answer, then—or a partial answer—be to permit children to be raised only by parents who really want them (assuming careful scrutiny of the reasons *why* they want them)? "People who are brought up without parental love," says A. S. Makarenko, the Soviet counterpart of America's Dr. Spock, "are often deformed people." And these deformed people are the ones who will make tomorrow's world. Makarenko, in his marriage manual, admonishes prospective parents: "If you wish to give birth to a citizen and do without parental love, then be so kind as to warn society that you wish to play such an underhanded trick." Presumably there would be enough volunteer parents who would gladly supply loving homes for such unwanted children.

When family units were larger—in older, less urbanized days—when there were grandparents in the house, or even aunts and uncles—a child had alternate sets of adults to turn to, with a wider chance of getting the kind of love and attention he needed, at the time he needed it. Even when there was not a large family living under one roof, people used to stay put longer in their communities, so that lots of relatives, and long-time friends who were almost like relatives, were in the immediate neighborhood during the years when a child was growing up. But now the typical family is a "nuclear" one, with only the married couple and their immediate children, living in a separate house or apartment. They probably have not lived there too long and may contemplate moving again soon. Chances are that no relatives live with them—or even close by—and that they are not too "involved" with their friends and neighbors, so the children are dependent for emotional sustenance solely on their single set of hopefully stable parents, and their human experience is thus considerably restricted. (If Mama wants to go out, she has to get a baby-sitter; she can't just go off, as she might with a larger household containing other adults.)

Among the peoples she has studied, Dr. Mead believes the Samoans are by far the best adjusted sexually and maritally. "When we examine how this capacity for reliable sex responses that nevertheless do not threaten or disrupt a social order which is firmly built into quite stable marriages, we find that the relationship between child and parent is early diffused over many adults . . . he is given food, consoled, carried about, by all the women of the large households, and later carried about the village by child-nurses who cluster together with their charges on their hips." But, except in our nostalgic fantasies, this type of large, tribal, multiparental household or community is a thing of the past in America. I suppose groups of friends or relatives (more likely friends; relatives seldom choose one another) could arbitrarily decide to live together in groups again, sharing households, expenses, and parental duties (just as neighbors now trade around baby-sitting chores on occasion). With several fathers around, a few of them might, like Arapesh males, enjoy spending lots of time with the kids and helping around the house, while the more restless and adventurous types would go forth to the marketplace. In Sweden today a few groups are actually trying this kind of experimental arrangement. Many of the hippies are certainly giving it a go. But could this form of tribalism work in our highly mobile, technological society?

Anthropologists in the last century—Briffault among them—thought that group marriage was quite common in earlier days and among primitive peoples. But this notion has been pretty well exploded in the twentieth century. The great comparative anthropologist, Dr. George P. Murdock, wrote that "there is no evidence that group marriage anywhere exists, or ever has existed, as the prevailing type of marital union." Dr. Malinowski certainly took a dim view of it and deplored as subversive "the idea that parenthood can be made collective." His clincher: "The only example of real group maternity I heard of was from a farmer friend of mine; he had three geese who decided to sit communally on a nest of eggs. The result was that all the eggs were smashed in the quarrels and fights of this maternal clan of group mothers. All, that is, but one; the gosling, however, did not survive the tender cares of its group mothers. If ever another group of geese were to try a similar experiment I should like them to be aware of this precedent."

Malinowski may very well be right. But he was talking in another social context, the context of 1931. Amid today's greater uncertainties, some people are seriously reconsidering the idea of group marriage. Farfetched though it may be, it is not to be dismissed out of hand.

In an era where permanent marriages were a rarity and children were raised only by volunteer parents, what would happen to the children when the parents separated? Whose children would they be? Would they be reassigned to some other group or couple for a while? Or, for stability's sake, would they have to be raised by the state—perhaps in small, family-like units?* Children could not even be assured continuity of locale as buildings became overcrowded or outdated, calling for reassignments of inmates to new quarters. We are familiar with the often ravaged personalities of orphan children who have gone from home to home. Can we imagine a whole population raised in this fashion? And would the full-time professional career child-raisers be any more competent at their jobs than most other civil servants?

And in this new picture, what would be the role of sex? If it were as casual as any other harmless pleasure (assuming the harmlessness of it), what would be wrong with anyone having sex with anyone else for no other reason than their mutual desire? Some people have been saying, in effect, "Good! It's about time sex were devalued and put in its place. Now maybe people will marry for more sensible reasons." But this kind of freedom could bring about a drastic decline in the quality of sexual experience—as well as a drastic reversal in the roles of both sexes.

Such a reversal would give neither sex much to rejoice at. Traditionally the male has been much freer about sex than the

---

* Utopian novelists have played with this idea, among them Dr. B. F. Skinner in *Walden Two* and Aldous Huxley in *Island*.

female. He was expected to delight in sex, to be the aggressor, the panting pursuer, the sower of wild oats. In the sex act, it was the woman who bestowed the favors, the man who won them. The woman treasured her chastity, used it as a lure to marriage. One of the reasons a man married was to assure himself secure possession of a pleasure that was otherwise hard to get. The woman submitted to his passion as her wifely duty.

Women have increasingly emancipated themselves from this mystique. They hear and read a great deal about their orgasms, what monumental experiences they can and should be, their inalienable right to have sexual pleasure in quantity and with frequency—yea, even into the seventies and eighties. We now know that the sexual needs of women are at least as great as those of men, and that the female climax is more intense and longer-lasting.

The male sexual capacity, despite the Casanovas and Don Juans of history, seems to be essentially more limited than the female's,* and his need more easily satisfied. It will be satisfied even more easily if the female goes on the prowl. He will not have to pursue at all. All he need do is stand there. Soon he may find himself fleeing as opportunities surpass his ability to deal with them. The woman, who formerly competed for males as marriage partners, may find herself competing for them as sexual partners. She will have become the aggressive pursuer of coy, hesitant males (even if the coyness and hesi-

---

* Writing about his male patients who are happy about the relative frigidity or passivity of their wives, psychoanalyst Milton R. Sapirstein says: "Underlying the strange preference is his awareness—rarely acknowledged, but existing just the same—of a basic biological fact, the fact that women have a far greater sexual potential than men. If he wakes the sleeping giantess, she may prove too much for him. Asleep or somnolent, on the other hand, she is no threat. The average healthy female—even in our own culture, where she is still relatively repressed—is capable of much more sexual activity than she can possibly experience with any one man."

tancy are due merely to satiation—or boredom with a commodity so totally available). Even today, as a woman grows older, she finds there are fewer and fewer men to go round. (For one thing, they tend to die earlier—those who do not defect to homosexuality at some point in their lives.) With so many enhanced opportunities for dissipation, the men may begin to wear out even sooner. To preserve them longer, women—especially if they can begin to have their babies without having to carry them, thus freeing them from their ancient bondage—may wind up working while the more delicate male stays home and takes care of himself. (Ah, but will he? Or, while she's away at work . . . the icelady cometh . . . or the Fuller Brush lady? . . .)

And as the supply of available males dwindled in a world where regular sex was every woman's right, what would women do? Would there be a return to polyandry? Would they simply practice wholesale automanipulation? Would they turn to each other? Would some of the ingenious artificial devices designed for current sex research go on the market—as well they might—so that any woman could get one as easily as she can now get a diaphragm? These devices are so skillfully contrived* that a woman who got addicted to them might prefer them to men. If that happened, perhaps the men—no longer pursued—would become the pursuers again.

However it all went, the concept of adultery would disappear, words like premarital and extramarital would become

---

* As described by researchers, one of the artificial penises designed for use by female subjects in the laboratory "can be adjusted for physical variations in size, weight and vaginal development. The rate and depth of penile thrust is initiated and controlled completely by the responding individual. As tension elevates, rapidity and depth of thrust are increased voluntarily, paralleling subject demand. The equipment is powered electrically." Less sophisticated vibrating devices are advertised as tension-relaxers in some of the pulp magazines.

meaningless, and no one would think of attaching a label like "promiscuity" to sex activities any more than to eating. After all, why not be as free to experiment with a variety of sexual partners as with a variety of foods and restaurants? Homosex could hardly be frowned upon under such circumstances. Masturbating would be like scratching an itch, and presumably perfectly all right with everyone, as long as the sperm was saved for the freezer.

Love, marriage, and the family have been around a long time, and have served us very well. But it is clear that they may not survive the new era unless we really want them to.

Whatever our attitude, a more liberalized sexuality does seem here to stay. I am glad to see it finally established—even among many churchmen—that sex is, or ought to be, a good and joyous thing.

"The time is past," Ardis Whitman quotes Dr. Gibson Winter as saying, "when the churches could hope to rule the sexual behavior of their members through laws and precepts." When Whitman asked Dr. Joseph Fletcher, the Episcopal theologian who wrote *Situation Ethics,* "Do you fear that these new ideas will lead to license?" Dr. Fletcher answered, "I do fear it. Sex is dynamite. In the past we have tried to guard ourselves against this dynamite through legalistic controls. Now we are going to have to look for positive rather than negative commandments; for devotion to an ideal rather than to the fear of consequences."

In this atmosphere, most enlightened authorities tend to agree with Dr. Fletcher's further judgment: "It is doubtful that love's cause is helped by any of the sex laws that try to dictate sexual practices for consenting adults." It looks very much as if we will have to abandon our old habit of insisting that sex must serve the same purpose for everyone; or even for the same person at different times of his life. As long as sex is practiced in private between fully consenting adults who do no physical harm to one another, it is simply not a matter for the police or for criminal statutes. The law must, of course, continue to protect children from any kind of sexual molestation, and adults—including prisoners—from any sexual acts involving force or coercion. Though sex *per se* may be morally neutral, what we do with it is not. It should never be used as a weapon, or as a means of exploiting others.

A good many authorities—among them Dr. Mead, Dr. Calderone, the Reverend Greet, and Masters and Johnson—have suggested that it might help if we stopped equating sex with mere intercourse; if we thought of sex as something a person *is* rather than something he *does*. A man's or woman's sex organs and sex acts should be incidental to his or her total sexuality—maleness or femaleness—as it develops from early childhood into old age, and as it fits.

How important should sex be in anyone's life? It depends on the individual and his circumstances. For people who are really in love, says Dr. Abraham H. Maslow of Brandeis University, sex is less important, yet at the same time more important, than it is for other people. The paradox is only apparent. Such individuals put sex in its proper place: Sex matters less, in the sense that it is neither compulsive nor obsessive, and its absence is tolerated with a fair degree of equanimity. (A person in love will of course feel deprived at the absence of the loved one—and sex is included in his sense of deprivation; but he does not miss sex-in-the-abstract, because the sex act, without love, is less likely to interest him.) Yet sex matters more to this person because, when it *is* practiced with his partner of choice, it provides an incomparable experience.

Maslow, who is current president of the American Psychological Association, has devoted many years of his research life to the study of psychologically healthy people. In discussing his "healthy" subjects, Maslow pointedly avoids the use of the word "normal"—since norm implies average. These people are not average. They are all too rare. Though no reliable statistics are available, Maslow would not be too surprised if they constituted less than 1% of the population. This may be why so many people seem skeptical that the heights of sexual ecstasy are ever reached outside of storybooks; it is possible, for all we know, that only a favored minority are equipped temperamentally to achieve it—though certainly more could than actually do. The members of this favored minority Maslow calls "self-actualizing" people, because they seem to be in the process of bringing into actuality their full human potential. Self-actualizers feel secure, they know who they are, and they think well of themselves. They are "loving as well as loved," and they are capable of doing both without fear or reservation.

When then, looked at in this light, is sex good, and when not? Dr. Fletcher offers a suggestion which I, for one, am happy to endorse: "Whether any form of sex . . . is good or evil depends on whether love is fully served."* At its most meaningful in human terms, sex is the unique means for expressing a mature love. But this is an old solution, not a new one. Yet I feel this is a solution we must come back to on a new level. In order for love to thrive, though, so must mature human relationships—which can only be between people able to communicate with one another, who care about one another, about the world, and about themselves. Caring about the self—caring, not in the sense of a mindless, selfish search for kicks, but a valuing of the self for its true human worth—is the first and perhaps the roughest hurdle. Sure knowledge of the self is never easy to attain. It never was, for any man, in any society,

---

* Recent studies of sex on the campus indicate that students have lately been tending increasingly to seek a loving relationship in which sex plays its part in preference to looking for the casual one-shot sex encounter. (But law in most states still punishes cohabitation much more severely than fornication.)

at any time in history. In these days when the underlying themes of our lives seem to be alienation and depersonalization, it is surely harder than Socrates ever thought it could be when he pronounced his celebrated injunction. Again, we can hope that the leisure afforded us in the new age will give us the time and the opportunity to come to terms with ourselves; or that the new knowledge will bring us the needed insights. We cannot of course overlook the possibility that the new knowledge, when it begins to be applied, will only confuse matters further for a while.

A man of our time who feels overburdened by his confusions— sexual and otherwise—and his responsibilities—including his marital ones—might see distinct advantage in the more care-free kind of world that the new biology could make feasible. On a bad day he might even envy his imaginary counterpart in one of the possible societies of the not-too-far-off future—a man grown *in vitro,* say, and raised by a state nursery. Such a man, it is true, might never know who his genetic parents were, nor would he have any brothers or sisters he could call his own. On the other hand, if he considered all men his brothers, what need would he have for a few specifically designated siblings who happened to be born in the same household? Think how carefree he might be: no parents to feel guilty about neglecting, no parental responsibilities of his own, no marriage partner to whom he owes fidelity—free to play, work, create, pursue his pleasures. In our current circumstances, the absence of a loved one saddens us, and death brings terrible grief. Think how easily the tears could be wiped away if there were no single "loved one" to miss that much—or if that loved one were readily replaceable by any of several others.*

---

* Taking a cool look at these possibilities, Gagnon and Simon are not at all sure the results would not be preferable to the current state of affairs.

And yet—if you (the hypothetical *in vitro* man) did not miss anyone very much, neither would anyone miss *you* very much. Your absence would cause little sadness, your death little grief. You too would be readily replaceable. A man needs to be needed. Who, in the new era, would need you? Would your mortality not weigh upon you even more heavily, though your life span were doubled or tripled?

"Which of us has known his brother?" wrote Thomas Wolfe. "Which of us has looked into his father's heart? Which of us has not remained forever prison-pent? Which of us is not forever a stranger and alone?"

The aloneness many of us feel on this earth is assuaged, more or less effectively, by the relationships we have with other human beings—our deep, abiding relationships with our parents, our children, our brothers and sisters, our wives, husbands, sweethearts, lovers, closest friends. These relationships are not always as deep or as abiding as we would like them to be, and communication is often distressingly difficult. Yet there are deep, full, loving relationships. And perhaps they are not as rare as the studies and charts would suggest. And there is always the hope that each man and woman who has not found such relationships will eventually find them. But in the *in vitro* world, or in the tissue-culture world, even the hope of deep, abiding relationships might be hard to sustain. Could society devise adequate substitutes? If each of us is "forever a stranger and alone" here and now, how much more strange, how much more alone, would one feel in a world where we belong to no one, and no one belongs to us. Could the trans-humans of post-civilization survive without love as we have known it in the institutions of marriage and family?

Suppose people *were* as replaceable as, say, things, clothing, houses, buildings, offices, occupations? Our joys would be less intense, but so would our frustrations and sorrows. In the absense of a strong sense of possessiveness, emotional attachments would not be so all-consuming, hence—in their view—there would probably be much less trouble in the world.

PART III

Control
of the
Brain
and
Behavior

IT IS EASY TO SEE WHY DESCARTES MADE "I THINK, THEREFORE I am" the baserock of his epistemology. A man's sense of individual identity is his concept of himself as a unique and relatively indestructible personality—what the scientist-philosopher Dr. Jacob Bronowski calls the self inside the skin, the self readily distinguishable from the world outside the skin, and from the other individual selves who can be recognized and related to in dependable ways. This awareness, the awareness of a distinct and separate self operating in a familiar physical and social milieu, exists nowhere if not in the mind. And that abstraction, the mind,* seems to be centered in the concrete physical entity of the brain—in the labyrinthine corrugations of the three pounds or so of "gray matter" inside the skull, in its geometry and its architecture, in the lavishly intricate electrochemistry of its ten billion intercommunicating nerve cells.

If the social personality is dependent upon the physical brain, it follows that changes in the brain will result in concomitant personality changes. This happens inadvertently when a man suffers, say, a stroke, or a severe head injury. If the brain damage is drastic, so are the personality changes likely to be. The thought has certainly occurred to people in the past—at least to scientists and to science-fiction writers—that the power to change the brain would provide the corollary power to change personality. Surgeons and psychiatrists have worked on this premise in using such crude, hit-or-miss techniques as prefrontal lobotomy and electroshock therapy in attempts to manage personality disorders considered to be otherwise unmanageable.

---

* "Among present-day philosophers," said the late Sir Russell Brain, "there is no agreement as to the proper use of the words mind and mental, but much of the confusion, which appears to be semantic in origin, might have been avoided if people had been content to speak in terms of human subjective experiences and behavior rather than of hypothetical entities."

The human personality can of course undergo quite radical changes through experiences that are altogether nonphysical in nature. These experiences can be sudden and spontaneous (religious conversion, poetic insight, mystic revelation), or gradual and deliberate (education, propaganda, brainwashing, persuasion of all kinds). Even in this kind of behavior transformation, the personality changes are probably accompanied by changes in the molecular structure and the electrochemical activity of some of the brain's cells. Thus, external circumstances, operating via the personality and mind, can bring about physical changes in the brain. As the noted neurophysiologist, Dr. Ralph W. Gerard, put it in his now-classic statement: "There can be no twisted thought without a twisted molecule."

This succinct and deliberately oversimplified comment sums up the daring new approach to brain research. The strong, though yet-to-be-altogether-proven, belief is that mind and matter, brain and behavior, are one. They can be thought of and dealt with separately as an artifice for man's convenience, but in reality they are inseparable.* In this scheme of things, changes in the mental sphere are never unaccompanied by changes in the physical. All our thoughts, feelings, moods, tastes, perceptions, attitudes—as they are formed in the slow growth and maturation of each of us—appear to lay down certain patterns in the cells of our brains. Researchers in hypnosis have been fascinated by the physical lesions (such as burns and blisters) that can be induced, and others (such as warts) that can be removed by pure suggestion to a subject

---

* Rather than words like "matter" and "mind," Sir Julian Huxley prefers the neutral German *Weltstoff,* or world stuff. "We then," he says, "are organizations of world stuff, but organizations with two aspects—a material aspect when looked at objectively from the outside, and a mental aspect when experienced subjectively from the inside. We are simultaneously and indissolubly both matter and mind."

under deep hypnosis. Physicians, increasingly aware of the psychosomatic nature of many ailments, observe every day how physiological changes can subtly influence the mental outlook of a patient.

The prospect arises, then, that we might learn to control human behavior by learning to change at will the patterns in our brain cells—by learning, as it were, to untwist the "twisted molecules." Untwisting could mean restructuring the cellular molecules in any way—rearranging their constituent atoms, taking away or adding a few ingredients, modifying the chemical or electrical activity. Ultimately chemistry and electricity come down to the same thing—the movements and rearrangements of electrons. Thus it is proper to think of the "electrochemistry" of the brain as an entity, or as a closely related series of processes and events that take place in the real four-dimensional physical world and which are therefore subject to both observation and manipulation. Manipulating the electrons of the brain, then, would equal manipulating human behavior.

Whether man can ever attain such a state of total detailed knowledge of the brain's workings, or the biotechnical sophistication needed to rescramble the electrons in the precisely calculated manner implied here, is doubtful. But man has already impressively demonstrated, by means of the so-called mind drugs, how powerfully the mind and personality can be affected by physical intervention; and, by means of electrical brain stimulation, that man can go a long way toward controlling behavior while possessing only sketchy and incomplete knowledge.

Studying behavior in a small monkey colony, for example, Dr. Jose M. R. Delgado of the Yale University Medical School, a pioneer researcher in this area of brain research, found that by remote radio stimulation of certain areas of the brain (where tiny electrodes had been surgically implanted), he could set in motion an entire sequence of activities. Dr. Delgado in a lecture at New York's Museum of Natural History described one facet of his work with an experimental monkey named Ludi:

> After different areas of the brain had been studied under restraint, the radio stimulator was strapped to Ludi, and excitations of the rostral part of the red nucleus were started, with the monkey free in the colony. Stimulation produced the following complex sequence of responses: (1) immediate interruption of spontaneous activities, (2) changes in facial expression, (3) head turning to the right, (4) standing on two feet, (5) circling to the right, (6) walking on two feet with perfect preservation of equilibrium by balancing the arms, touching the walls of the cage, or grasping the swings, (7) climbing a pole on the back wall of the cage, (8) descending to the floor, (9) low tone vocalization, (10) threatening attitude toward subordinate monkeys, (11) changing of attitude and peacefully approaching some other members of the colony, and (12) resumption of the activity interrupted by the stimulation.

The detailed knowledge that Delgado would have needed to make a monkey go through all this, starting from scratch, would be phenomenal, and certainly far beyond anyone's present grasp. But by taking advantage of patterns already there, as if preprogrammed in a computer, the mere stimulation of the right area of the brain can set the entire sequence in motion.

When you press a light switch, it is not the flick of the switch that turns on the light. It merely acts as a trigger to set in motion the chain of events—the flow of electrons through wires, the glowing of the tungsten filament, and so on—that results in illumination. But by knowing what to do, you do in fact control the light. In the same manner, when a space-flight controller at Cape Kennedy or in Houston finishes a countdown, it is not the sound of his voice or the pressing of a button that launches the astronauts into orbit. Yet, by triggering—or deciding not to trigger—this preprogrammed series of events, he does control the outcome. Similarly Dr. Delgado learned, without direct detailed interference in the submicroscopic events in the monkey's brain, to control the behavior sequence by controlling the trigger.

> The whole sequence was repeated again and again, as many times as the red nucleus was stimulated. Responses 1 to 8 developed during the five seconds of stimulation and were followed, as aftereffects, by responses 9 to 12 which lasted from five to ten seconds. The excitations were repeated every minute for one hour, and results were highly consistent on different days. The responses resembled spontaneous activities, were well organized, and always maintained the described sequence. Climbing followed but never preceded turning the body; vocalization followed but never preceded walking on two feet; the general pattern was similar in different stimulations, but the details of motor performance varied and were adjusted for existing circumstances. For example, if the stimulation surprised the animal with one arm around the vertical pole in the cage, the first part of the evoked response was to withdraw the arm in order to make the turn possible. While walking on two feet, the monkey was well oriented and was able to avoid obstacles in its path and to react according to the social situation. In some experiments, three monkeys were simultaneously radio-stimulated in the red nucleus, and all three performed the full behavioral sequence without interfering with one another.

The stimulus always worked unless it was overridden by a more powerful set of demands. A monkey eating after it had starved for twenty-four hours, for example, or a monkey threatened with physical danger, tended to pay attention to its hunger or its self-defense rather than be overwhelmed by the stimulation of the red nucleus. But in general, under normal circumstances, when not under any great pressures or threats to its life, the monkey performed as dictated to. "Examples of other patterns of sequential behavior," says Delgado, "have been evoked by excitation of several diencephalic and mesencephalic structures, showing that sequential activities are anatomically represented in several parts of the central nervous system." An understanding of the brain's geographical terminology is not necessary to the understanding of the implications of Delgado's findings.

Take a look, for instance, at another of Delgado's experiments, this one in a different monkey colony, where patterns of sexual behavior were triggered by the same kind of direct and simple stimulus.

> Radio stimulation of the nucleus medialis dorsalis of the thalamus in a female monkey produced a sequential pattern of behavior characterized by a movement of the head, walking on all fours, jumping to the back wall of the cage for two or three seconds, jumping down to the floor, and walking back to the starting point. At this moment, she was approached by the boss of the colony, and she stood on all fours, raised her tail and was grasped and mounted by the boss in a manner indistinguishable from spontaneous mounting. The entire behavioral sequence was repeated once every minute following each stimulation, and a total of 81 mountings was recorded in a 90-minute period, while no other mountings were recorded on the same day. As is natural in social interaction, the evoked responses affected not only the animal with the cerebral electrodes, but also other members of the colony.

The probability that this much sexual activity would have taken place over this period of time in the normal course of events, without the electrical command from outside, is negligibly low. Yet, in each of these behavioral sequences—sexual or otherwise—once the series of events was set in motion, it then proceeded exactly as if the whole thing had been the monkey's idea in the first place. It seems likely that, when this kind of stimulated sequential behavior takes place, the monkey believes it *is* her own idea. This likelihood has been borne out in experiments with human beings. The subject has no *feeling* that his brain cells are being electrically stimulated—and would not know it was being done if the experimenter chose not to tell him. He only experiences the resulting sensations as if they had come normally and spontaneously.

ESB (not to be confused with ESP—for extrasensory perception; or BSP—for biosocioprolepsis) stands for electrical stimulation of the brain. ESB, the technique used by Delgado in his monkey experiments, stems originally from brilliant work done with cats back in the early 1930's by the Swiss Nobel laureate, Dr. Walter R. Hess. "The principle of ESB," wrote Robert Coughlan in *Life,* "is simple: stick an electrical conductor into whichever part of the brain one happens to be curious about, turn on the current and see what happens."

The usual laboratory method, as described by Coughlan, is this: "The head of the anesthetized subject is immobilized. A tiny high-speed drill is used to bore through the skull and sink a minute well shaft through intervening tissue to the point chosen for investigation. With a micromanipulator the operator then inserts an assembly consisting of a miniature electrode attached to two insulated wire filaments. The other, or scalp-side, ends of these wires are then connected with a small terminal socket and the latter is cemented to the skull. Current fed into the socket goes down the pair of filaments to the electrode and supplies the stimulus—as if the nerve cells there had all fired electrical discharges in unison.

"So incredibly exact have this technique and apparatus become," says Coughlan, "that a microelectrode only a millionth of an inch in diameter can be placed *inside* an individual nerve cell . . . without interfering with any of the cell's normal processes.

"Incidentally," he adds, "this procedure does not hurt the subject in the least. Most parts of the brain, oddly enough, are not able to feel pain, and the electric current is kept at low levels." (It is this lack of pain in brain surgery that enables the wide-awake patient—a victim, perhaps, of epilepsy or Parkinson's disease, to describe his reactions to the surgeon as various manipulations are in progress. Often the surgeon could not perform his task without the patient's verbal assistance to guide him.)

Experimental animals soon get used to their skull sockets, and, according to Delgado, "extensive experimentation by many authors [i.e., authors of scientific papers] has demonstrated that intracerebral electrodes are safe and can be tolerated for years, providing an effective tool for sending and recording electrical impulses to and from the brain of unanesthetized animals." Delgado himself has experimented with cats, dogs, mice, monkeys, and bulls. Others—among them Dr. Carl W. Sem-Jacobsen in Norway, Dr. Robert G. Heath at Tulane, and Drs. Vernon Mark and William Sweet at Harvard—have gone on to implant electrodes in the brains of human beings. Obviously this has not been done casually, and the practice has been restricted to people who were patients rather than experimental subjects—patients whose ailments (epilepsy, intractable pain, anxiety neurosis, involuntary movement) could be helped by these techniques. "Accumulated experience," says Delgado, "has shown that electrodes are well tolerated by the human brain for at least one year and a half, and that electrical stimulations may induce a variety of responses, including changes in mental function. . . . The prospect of leaving wires inside the thinking brain could seem barbaric, uncomfortable, and dangerous, but actually the patients who have undergone this experience have had no ill effects, and they have not been concerned about the idea of being wired or by the existence of leads in their heads. In some cases, they enjoyed a normal life as outpatients, returning to the clinic for periodic stimulations. Some of the women proved the adaptability of the feminine spirit to all situations by designing pretty hats to conceal their electrical headgear."

Control of the brain would also provide control of those primitive areas which in turn control the basic functions of the body. "Buried at the base of the brain," Dr. Joel Elkes of Johns Hopkins points out, "in the midline, the center of the head, there are old, old regions, concerned clearly with survival. These areas control respiration, pulse rate, and blood pressure; govern salt balance and temperature control; guide certain built-in instinctual responses such as hunger, thirst, fight, flight, play, sleep, wakefulness, sex. These are the steering centers of the cerebral machinery."

ESB experiments have indeed already shown that an animal can be induced to starve itself (though it has gone hungry for some time) or gorge itself (though it has just eaten), or to perform sexually far beyond its normal capacity. The kinds of controls that can be exerted on animals and men by ESB range all the way from simple muscular movements to fairly complex social behavior. It has been known at least since the nineteenth century that electrical stimulation of the cerebral cortex could produce motor responses in animals. But until recently it was assumed that this could be achieved only with anesthetized animals, and that the movements would be clumsy and imprecise. But the newer techniques and miniaturized apparatus that made ESB possible have made it evident, as Delgado says, "that motor performance under electronic command could be as complex and precise as spontaneous behavior."

Delgado describes an induced leg movement—the flexing of a hind leg—in a laboratory cat as an example:

> The evoked movement usually began slowly, developed smoothly, reached its peak in about two seconds, and lasted until the end of the stimulation. This motor performance could be repeated as many times as desired, and it was ac-

companied by a postural adjustment of the whole body which included a lowering of the head, raising of the pelvis, and a slight weight shift to the left in order to maintain equilibrium on only three legs.

Did all of these ESB commands disturb the cat emotionally? On the contrary:

> The cat was as alert and friendly as usual, rubbing its head against the experimenter, seeking to be petted, and purring. However, if we tried to prevent the evoked effect by holding the left hind legs with our hands, the cat stopped purring, struggled to get free, and shook its leg. Apparently the evoked motility was not unpleasant, but attempts to prevent it were disturbing for the animal.

These reactions are not dissimilar from those of humans under ESB. This being the case, Delgado's further comment on the cat experiment is particularly to be noted:

> The artificial driving of motor activities was accepted in such a natural way by the animal that often there was spontaneous initiative to cooperate with the electrical command.* For example, during a moment of precarious balance when all paws were close together, stimulation produced first a postural adjustment and the cat spread its forelegs to achieve equilibrium by shifting its body weight to the right, and only after this delay did the left hind leg begin to flex. . . . A

---

* Dr. Gregory Razran of Queens College, New York, who has made a continuing study of Russian psychology one of his specialties, says that in the U.S.S.R., experimental psychologists of the Pavlovian persuasion have made a careful distinction between conditioned reflexes (e.g., salivation) that are triggered from the outside (e.g., the ringing of a bell), and those triggered by stimulation of an internal organ (e.g., the bladder). Reflexes triggered by internal stimulation, says Razran, are always much more unconscious. It is likely that ESB would fall in this category. Though the triggerer is on the outside, the stimulation occurs inside and appears to be indistinguishable from the natural occurrence of an idea.

variety of motor effects have been evoked in different species, including cat, dog, bull, and monkey. The animals could be induced to move the legs, raise or lower the body, open or close the mouth, walk or lie still, turn around, and perform a variety of responses with predictable reliability, *as if they were electronic toys under human control.* [italics mine].

Moreover, animals seem to enjoy being stimulated electrically—another disquieting phenomenon if translatable to people.

Going beyond these simple motor activities and the more complex sequences of activities described earlier, Delgado and others found they could also affect moods, attitudes, and even the basic character of individual animals (which in turn affected the behavior of other animals) by stimulating the appropriate points or regions of the brain. A cat can be induced to start a fight with another cat—or a dog—much larger than itself; or to cringe from a mouse, depending on the brain area getting the signals. A peaceful animal can be made to snarl and turn belligerent, while a normally aggressive animal can be rendered docile. Rhesus monkeys, says Delgado, "are destructive and dangerous creatures which do not hesitate to bite anything within reach, including leads, instrumentation, and occasionally the experimenter's hands. Would it be possible to tame these ferocious animals by means of electrical stimulation? To investigate the question, a monkey was strapped to a chair where it made faces and threatened the investigator until the rostral part of the caudate nucleus was electrically stimulated. At this moment, the monkey lost its aggressive expression and did not try to grab or bite the experimenter, who could safely put a finger in its mouth! As soon as stimulation was discontinued, the monkey was as aggressive as before.

"Later," Delgado goes on, "similar experiments were repeated with the monkeys free inside the colony, and it was evident that their autocratic social structure could be manipulated by radio stimulation." The boss monkey, under ESB, lost his aggressiveness, and the other monkeys crowded him without fear. This went on for about an hour. "About 12 minutes after the stimulation hour ended, the boss had reasserted his authority." In similar experiments at the Yerkes Regional Primate Center, Dr. Bryan W. Robinson noted that, as first one and then another male became dominant, "the female switched her allegiance to the dominant male, and then turned about and attacked the other guy." In an even more interesting version of the experiment, Delgado observed that the other monkeys in the colony "learned to press a lever in the cage which triggered stimulation of the boss monkey in the caudate nucleus, inhibiting his aggressive behavior." Thus one monkey was deliberately controlling the behavior of another by means of ESB—a truly impressive demonstration of how little needs to be understood to exercise quite a lot of control.

Perhaps the most dramatic demonstration of what ESB could do in the way of turning off aggression—and how confident a scientist could be of ESB's power—was a bravura performance by Delgado himself with a real fighting bull (into whose brain he had implanted electrodes) in a Spanish bull ring. Playing the role of matador, the cape-waving Delgado got the animal all worked up to the proper pitch of snorting, pawing ferocity. Then, standing there calmly as the bull charged, he stopped the bull within a few feet of him. He had literally turned off the bull's charge by means of a small radio transmitter he carried in his hand. ESB had rendered the bull as friendly as Ferdinand.

Often the stimulated area that makes an animal very angry is located quite nearby the area that makes it euphoric. An electrode implanted at one spot in the amygdala might, when ESB is applied, bring on a paroxysm of ungovernable rage; if it is moved only a fraction of an inch away, ESB will result in the most friendly, even loving, behavior. Could this produce a world where no one would ever be angry? Would this be a good thing?

This is not an idle question. The uses of ESB in the control of *human* aggression has already been convincingly demonstrated at a clinic in Boston which makes a specialty of studying violent behavior. It was organized in 1967 by a team of medical scientists attached to Harvard Medical School, Massachusetts General Hospital and Boston City Hospital. The group includes two outstanding brain surgeons, Dr. Mark and Dr. Sweet, and its full-time head is a psychiatrist, Dr. Frank R. Ervin.

The typical clinic patient has "poor impulse control" with a quick-flaring temper and a history of repeated violent episodes. Many of these patients are incredibly destructive of property, and they may beat their wives, husbands or children with astonishing ferocity. One young wife who came in recently for help said that she had assaulted her husband—fortunately a very large, very tolerant man—537 times in the last six years, with everything from fists to dishes to furniture. Violent people also frequently vent their impulses through sexual assault or multiple automobile accidents.

Though the violent patient usually has a "reason" for his uncontrollable rages, the reason can be incredibly flimsy: he may do major violence in response to a minor or imagined slight. A man may knock his wife across the room because she burned the toast. A teen-age girl may smash her room into a total shambles because her brother asked her to turn down the record player. Yet, between bouts of violence, this same person may be mild-mannered, charming and altogether likable. Once the rage is gone and the damage done, there may be a flood of guilt and contrition, sometimes followed by a near-suicidal depression.

In the 50-or-so cases the clinic has so far had the opportunity to study in some depth, there has been a startling frequency of correlation between deviant behavior and brain damage. The damaged or abnormal areas can often be pinpointed through ESB—and the rage evoked at will by stimulation, and, depending on the site and extent of the abnormality, turned off by ESB as well. In rare cases, because of the ravages accompanying certain types of epilepsy or the presence of a tumor in the primitive brain, the damage is so extensive that the patient is violent nearly all the time. The damage apparently scrambles the electrical circuitry so that the cells in the affected regions are discharging electricity almost constantly, evoking impulses of rage and violence. There is no way to turn them off, except through drug therapy or brain surgery.

So far there has been great reluctance to perform brain surgery, except in extreme cases—repeated attempts at murder, for instance. Sometimes even relatively simple surgery—if any brain surgery can be called simple—can help for a time. At the Indiana University Medical Center, Dr. Robert Heimberger has found that by touching the afflicted area of the brain with a delicate "cryosurgical" probe (an instrument with a frozen tip) he can destroy the diseased tissue. This operation, performed on institutionalized patients who are violently destructive, keeps them calm for weeks or months at a time.

In many of the cases handled by Dr. Sweet and Dr. Mark, the brain damage is not obvious. But examinations in depth usually turn up some abnormality in the tissue—damage that is perhaps congenital, perhaps the result of blows on the head, or of some viral infection that reached the brain. There has lately been much interest in genetic causes of these abnormalities, too, especially since a recent case in France, where a violent criminal was found to possess an abnormal "XYY" chromosome. The Boston group has already incorporated a cell geneticist into the team to study these latest possibilities.

All this obviously has important implications for criminology and penology. When I earlier cited the imaginary example of a rapist being cured of his tendencies, transformed by brain surgery from a sadistic brute into a gentleman of sweet disposition, it probably seemed far-fetched. But we can now see that the possibility may be more immediate than anyone imagined. Early in 1968, a British court handed over a young incorrigible—a "compulsive gambler"—to doctors for treatment by a leucotomy operation. After the story appeared in the London *Times,* the *British Medical Journal* expressed its qualms about what this sort of procedure might lead to in terms of sentencing criminals to treatment instead of to jail. Yet the precedent is already established: In cases of "insanity," the criminal is often turned over for psychiatric treatment rather than sentenced to prison (though he may spend an equally long time in confinement) on the grounds that it was his mental illness rather than the man himself that was at fault. Will the presence of brain damage or a bad chromosome soon be sufficient to absolve a criminal of guilt on the same grounds?

Of all ESB experiments carried out with animals, perhaps none was more astonishing than the series back in 1953 and 1954 in which Dr. James Olds (then at McGill University in Montreal) accidentally discovered the brain's pleasure centers. He had just learned how to implant electrodes in rat brains preparatory to studying rat behavior. But he was curious to know if the ESB technique itself might so disturb and distract the rats as to spoil his experiment. "I went up to the lab one Sunday afternoon," he recalls, "and took the first rat I had ever prepared with my own hands. Every time the rat walked into one corner of the testing table, I turned on the electricity to see if he would avoid approaching that spot thereafter. Instead, my rat *liked* it!"

Pursuing this windfall instead of his original idea, Olds refined his techniques, found that he could, by ESB, produce at will a state of bliss in the rat. Other researchers eagerly fol-

lowed Olds's lead and confirmed that there were a number of
pleasure centers in a variety of animals. Coughlan writes:

> Many sites seem to be identified with specific pleasures, such
> as those of food, drink, and sex. But sometimes ESB sets up
> a complex, generalized response. This may indicate a higher
> satisfaction independent of specific pleasures—or possibly,
> Dr. Olds suspects, that particular pleasure sites are packed
> close together and several are stimulated by one large dose
> of ESB.
>
> The nature of this feeling of pleasure is scarcely definable:
> some have guessed that it combines the mystical raptures of
> the saints with the fleshly raptures of the sinners, in a dif-
> fused, ineffable delight. In any case, animals find the sensa-
> tion completely irresistible. A white rat, if allowed to regu-
> late its own ESB dosage by pressing a lever in its cage, will
> continue to press it—at the fantastic rate of up to 8,000 times
> an hour—until hunger, thirst, or exhaustion force an inter-
> ruption. But the interruption is brief: a sip, a bite or two, a
> few minutes' nap, and the rat returns to its orgy of pleasures.
> In an experiment by Dr. Joseph V. Brady at Walter Reed
> Army Medical Center, rats went on this way 24 hours a day
> for three weeks straight. One would expect that such
> sybaritic rats would eventually wear themselves to a frazzle,
> burn themselves out before they were 30 days old, but to the
> contrary the ones at Walter Reed showed no physical or
> mental damage then or later. And Dr. Olds' rats, after a
> series of ESB marathons cumulatively totaling hundreds of
> days, have seemed in better health and fettle than their
> littermates who were raised in identical conditions but with-
> out ESB.
>
> Pleasure centers have been located also in cats, dogs, mon-
> keys, apes, and bottle-nosed dolphins . . . and these crea-
> tures have responded to ESB in the same degree.

Later experiments by Dr. Heath, by Dr. Sem-Jacobsen, and
again by Delgado indicate that the human brain, too, possesses
pleasure centers. Patients under ESB, sometimes without knowing that ESB had been applied at a given moment, suddenly
said they were experiencing highly pleasurable sensations. ESB

was able, at times, to turn depression to gaiety, and lethargy to alertness. Shy people became suddenly bright and talkative, and normally reserved women grew languorously flirtatious. Some of Dr. Heath's mental and epileptic patients have worn electrodes for long periods of time—electrodes they could themselves stimulate at will. This technique is called ICSS—for intracranial self-stimulation. ICSS devices have varied uses. A certain type of epileptic, for instance, feeling the first beginning sign of a convulsive seizure, can stop it instantly by pushing the button. A man afflicted with narcolepsy (chronic sleepiness) can stimulate himself into a state of wakefulness. In one patient with severe narcolepsy, the method worked so well that "by virtue of his ability to control symptoms with the stimulator," says Dr. Heath, "he was employed part-time, while wearing the unit, as an entertainer in a night club." This patient, like some others, had more than one button on his stimulator and had access to more than one area of his brain. He found that when he pushed one of the buttons "the feeling was 'good'; it was as if he were building up to a sexual orgasm." He pushed it frequently. So did another patient. On checking, Dr. Heath found that "regardless of his emotional state and the subject under discussion in the room," stimulation in this area "was accompanied by the patient's introduction of a sexual subject, usually with a broad grin. When questioned about this, he would say, 'I don't know why that came to mind—I just happened to think of it.'"

One can easily imagine people in the future wearing self-stimulating electrodes (it might even become the "in" thing to do) which might render the wearer sexually potent at any time; that might put him to sleep or keep him awake, according to his need; that might curb his appetite if he wanted to lose weight; that might relieve him of pain; that might give him courage when he was fearful, or render him tranquil when he was enraged.

The notion of a man controlling his own brain is one thing. But the prospect that a man's brain might be controlled by another man is something else again—not to mention the control of masses of people by a few powerful individuals. Delgado, for one, does not take this latter possibility too seriously. He admits that—through such practices as requiring blood tests before marriage, compulsory smallpox vaccination, and fluoridation and chlorination of our drinking water—governments "have established a precedent of official manipulation of our personal biology." He sees, too, that "governments could try to control general behavior or to increase the happiness of citizens by electronically influencing their brains." But, "fortunately," he concludes, "this prospect is remote, if not impossible, not only for obvious ethical reasons, but also because of its impracticability. Theoretically it would be possible to regulate aggressiveness, productivity, or sleep by means of electrodes implanted in the brain, but this technique requires specialized knowledge, refined skills, and a detailed and complex exploration in each individual, because of the existence of anatomical and physiological variability. The feasibility of mass control of behavior by brain stimulation is very unlikely, and the application of intracerebral electrodes in man will probably remain highly individualized and restricted to medical practice."

But not all scientists share Delgado's optimism about the remoteness of these prospects, especially since ESB is only in its infancy. The fact that it might be difficult or troublesome (and it could soon become less difficult and less troublesome) to apply ESB on a large scale would not necessarily deter someone who was sufficiently motivated to do it, and had the power to carry out his will. As for the "obvious ethical reasons," that would depend upon the individual ethics of the persons in power—and would of course carry no weight at all with ruthless types who invent their ethics as they go. It has been suggested that a dictator might even implant electrodes in the brains of infants a few months after birth—and they would never know that their thoughts, moods, feelings, and all-around behavior were not the results of their own volition. Where then is free will, and individual responsibility? An electrical engineer named Curtiss R. Schafer, who made a similar suggestion, added that "the once-human being thus controlled would be the cheapest of machines to create and operate. The cost of building even a simple robot like the Westinghouse mechanical man is probably 10 times that of bearing and raising a child to the age of 16."

"What does this imply?" asks Robert Coughlan. "One hypothetical possibility . . .: the 100-socket, 600-electrode human being controlled by a transistor-timed stimulator worn perhaps, in the form of a lapel pin by men and of a jeweled brooch by women. Each individual's program would be pre-set and tailored to assigned functions and duties, but it could be changed instantly by overriding radio signals sent out by local (75-socket) controllers, who would be controlled by district (50-socket) controllers, who would be controlled by regional (25-socket) controllers, who would be controlled by a Master Controller (no sockets) who, in his wisdom, would control the behavior of everybody."

So much for the electrical half of the brain's electrochemistry. But the chemical half may turn out to be considerably more than half in terms of its fundamental functional importance. Not that electrochemistry is neatly divisible into halves. But it is feasible to think of the electricity and the chemistry as separate though interrelated modes of cerebral operation; and certainly to think of electricity (e.g., the application of ESB) and chemistry (e.g., the administration of drugs) as two distinct approaches to the control of the brain. Potent as the electrical approach appears to be, chemistry looks even more promising, hence even more threatening.

Though neural impulses can be electrically stimulated, their actual transmission along the nerve fibers and across the synapses is achieved—as was demonstrated by Sir John C. Eccles and his colleagues at the Australian National University—by the transport of key chemical substances,* the principal transmitter in the central nervous system (CNS) being acetylcholine. And while thoughts and memories may go round and round in their electrical circuits, they appear to be stored chemically in the molecules of the brain—with implications for memory and learning to be discussed shortly. It has already been amply demonstrated that electricity introduced from the outside can turn on the chemistry in a given area of the brain and stimulate a variety of behavior patterns. In such cases it is the impingement of the electrical current that sets in motion the chemical reactions. Under normal circumstances, though, it works the other way around. The brain's cells are like miniature storage batteries, with an electrical potential that lies latent, stored chemically as positive or negative ions (atoms or molecules with extra electrons or missing electrons), until an appropriate stimulus sparks the electricity into action. This firing of the cells sets in motion the electrical currents that keep the brain and CNS functioning—much as the storage battery in your car, once it is turned on, initiates the current that runs the motor. These weak electrical currents are what the electroencephalograph picks up; it is when they are no longer detectable that most physicians are now willing to consider a patient truly dead.

If the use of a broadside technique like ESB can give its possessor such powers over the brain, think what might be

---

* One group of deadly "nerve gases" is known as *cholinesterase inhibitors*. By interfering with the activity of a single enzyme, cholinesterase, they effectively block a vital chemical cycle, thus turning off the nerve cells' chemical transmitting apparatus. When the nerve impulses stop, so does the heartbeat.

done if we knew how to manipulate the brain's chemistry in all its exquisitely refined detail. "Just as the DNA code determines the color of the eye, the shape of the nose, and the precise operations of such complex organs as the liver," writes Lawrence Lessing in *Fortune,* "so it also determines the cast of the mind. The new hypothesis is that DNA not only specifies the physical structure of the brain, but it also controls, directly or indirectly, all brain processes and mental activity through a molecular code that may be searched out and finally mastered."

Dr. Joel Elkes adds: "If certain behaviors are genetically coded, then these behaviors can be chemically released." Though ultimate mastery of these chemical codes is a real hope, we are now only at the bare beginnings of the necessary knowledge. In the words of Dr. Robert S. deRopp, "The scientist who attempts to study the chemistry of thought and feeling resembles a burglar attempting to open a vault of one of the world's major banks with a toothpick."

But an expert safecracker might, by thorough familiarity with the vault, figure out a way to do the job with a toothpick. There are events in science which the late Nobel chemist Dr. Irving Langmuir liked to call divergent phenomena—events which, though tiny in themselves, when applied at the right place at the right time, can start a chain reaction of happenings out of all proportion to the triggering action. A favorite example of Langmuir's was the series of occurrences in a Wilson cloud chamber, where a single quantum event, such as the disintegration of a radium atom, instantaneously produces thousands upon thousands of water droplets. This understanding of divergent phenomena is what gave Langmuir, one of the original "artificial rainmakers," the courage to try to modify massive weather patterns by tampering in very small ways.

A few stray neutrons in a critical mass of refined uranium can set off a nuclear blast. A sudden cry can cause an avalanche, and the slippage of rocks along a fault can result in a major earthquake. So it should come as no surprise that a little bit of interference with the chemistry of a few key cells in some tiny areas of the brain can be a divergent phenomenon, triggering major biological events. Consider, for instance, the rainbow visions, the ascents and descents into private psychic hells and heavens, the telescoping of time, the bizarre and long-lasting inner experiences that are evoked by one twenty-thousandth of an ounce of LSD—an amount too small to be seen of a tasteless, colorless, odorless substance. This is perhaps even more amazing than opening a bank vault with a toothpick.

At the base of the cerebrum there is a segment of the primitive brain called the hypothalamus which has a major role in a number of basic physiological functions. It has a role in sex, and in the sleep-wake cycle. It serves as the body's thermostat, its temperature regulator, as well as its appestat, or appetite regulator. If a man's appestat is off, he may eat too much and get too fat—or have no yen for food at all and waste away. ESB experiments have shown that when an animal's appestat is deliberately thrown out of kilter, the chemical signals it gets from its body as to its state of hunger or satiety become unreliable. So do the instructions it sends out in response.

By singling out the crucial cells in the hypothalamus, nearly all its functions can be altered quite radically. At the centennial celebration of the National Academy of Sciences, Dr. Neal E. Miller of Yale told of a series of experiments in which a thermode had been used, not to tamper with the basic chemistry of cells, but merely to heat or cool a tiny specific region of the anterior hypothalamus. The results were recorded with microelectrodes. Most of the nerve cells were found to be relatively unaffected by moderate temperatures changes. Miller said:

> However there are some neurons here that increase their rate
> of firing when they are slightly heated, and others that in-

crease their rate when they are slightly cooled. These cells seem to serve as, or be connected to, specialized "sense organs" for measuring small changes in the temperature of the surrounding blood.

Heating this region of an animal's brain causes panting and increased blood supply to the skin which serves to lower the body temperature. Cooling it causes the opposite effect of shivering and decreased blood supply to the skin. It also stimulates the secretion of the thyroid, which in turn speeds up the body's burning of fuel. In the experiments in which only this tiny region of the brain is cooled, these effects produce a fever, but when the whole body is cooled under normal conditions, they serve to restore the animal's temperature to normal. . . . Cooling this region of the brain will make a satiated animal hungry, so that it will eat, whereas heating this region elicits drinking. Thus, this temperature-regulating mechanism is tied in with hunger and thirst which motivate behavior that helps the animal to anticipate its need for the fuel it will burn to keep warm, or the water it will evaporate to cool off.

In short, a whole series of homeostatic mechanisms ranging from changes in metabolism to the motivation of the behavior of seeking food or water is touched off by the cells in the brain that respond to temperature.

Again a divergent phenomenon—a toothpick's worth of tampering produces a bank vault's worth of payoff. There is obviously great potential here for long-range medical and psychiatric benefits in the possibility of controlling body temperature, appetite, sleep, and sexual desire.

Still short of manipulating the cells' chemistry, there are yet other simple methods of achieving the same results as those obtained by heating and cooling. "Certain receptors in the brain," explains Dr. Miller, "respond to osmotic pressure so that a minute injection into the proper place in the brain of a solution that is slightly more salty than body fluid will motivate animals that have just been satiated on water to drink, and also to perform responses that they have learned to get water. . . . Conversely, a minute injection of water will cause a dehydrated animal to stop drinking or working for water."

Scientists have also now succeeded in going the extra steps to direct chemical interference. Studies by Dr. S. P. Grossman at Yale have shown, again in Dr. Miller's words, that "after a rat has been thoroughly satiated on both food and water, injecting a minute amount of acetylcholine or carbachol . . . will cause it to drink, while epinephrine or norepinephrine injected into the same site will cause the same satiated rat to eat." This experiment and other similar studies serve as clinching evidence for Miller "that the neuromechanisms involved in the motivations of hunger and thirst are chemically coded." Miller goes on to describe, too, how "yet other cells of the brain respond to specific hormones so that activities such as nest-building in rats can be elicited by injecting a minute quantity of the proper hormone into the correct site in the brain."

The effects of hormones used this way can be impressive indeed, as Dr. Elkes made clear in a Deerfield Foundation lecture in 1965. "A small dose of hormone, a steroid entering the central nervous system during an acute developmental phase," he said, "will so re-set its responsiveness that the whole program of behavior is one of male rather than female activity. Let me cite another experiment, coming from our own laboratory [at Johns Hopkins]. This was done by Dr. Richard Michael, and concerns adult cats. Dr. Michael implanted a minute quantity of hormone directly into a small area of the posterior hypothalamus. . . . He also implanted, in other animals, dummy material of the same dimension. . . . The cats implanted with small quantities of the hormone (silbesterol dibutyrate), although originally devoid of sexual activity, became sexually very receptive; dummy implanted cats did not show any such effects. With removal of the implants, the susceptibility disappeared." The Johns Hopkins researchers also found that the hormone does not travel very far, that it stays in the immediate vicinity of the implant, and that a few selected cells take up the hormone. This," says Elkes, "is a remarkable instance of specificity of interaction in the CNS."

"Let us think this to the end," he then implores. "A small amount of hormone incorporated properly into the membrane of relatively few cells so re-sets the total machinery that it now responds to a male presence with a very specific sexual response—a program that runs down in about eight minutes or so in a cat. Put in another hormone, and this will not happen. Move the hormone a few millimeters away from the susceptible site and again this will not happen. Cells are thus apparently sensitized in a highly specific way."

So a surprising lot is getting to be known. "We now have methods available," says Elkes, "which enable us to map this chemistry of the brain in great detail, not only in terms of gross, macroscopic structure—shall we say, the geography of the brain itself—but also in terms of the layer by layer geology of the brain; and to determine the concentration of those materials in very thin areas of the brain and show how one area differs from another, only a few millionths of an inch apart. It can be done by elegant microtechniques which enable one to gauge the local concentration of materials; it can also be done by special staining methods which show up these materials in beautiful florescent arrays."

It is quite clear, then, that the brain can be controlled chemically in at least a limited fashion. And, since all the behavior patterns capable of being set in motion by ESB are based on chemical patterns that are stored and ready-to-go, and are chemically carried out, it follows that anything ESB can do, chemistry can do better—once we learn how. At its best, ESB still requires the implantation of electrodes inside the brain and the cementing of sockets onto the skull, an exacting task whose end result is a relatively gross prodding of an area or site of the

213

brain. Chemical control—interacting directly with the substances in the brain cells without physical molestation, without destruction of tissue, without the necessity for electrodes or sockets—is obviously the preferred method, and would provide much more precise control, not only triggering behavior, but modifying behavior as well, and modifying it virtually at will.

The required sophistication is still a long way off. In the experiments described by Dr. Miller and Dr. Elkes, the chemicals were applied directly to the cells of the brains—but only, as in ESB, by means of microsurgical techniques. "Some of this work," says Miller, "has been done by biophysicists who thrust micropipettes with several barrels into a single nerve cell, using a conductive solution in one pipette to record the electrical activity of the cell, while minute quantities are injected electrophoretically via the other barrels. Studies with the electron microscope have verified other details. Yet other studies have used a push-pull cannula to wash out and measure for a group of nerve cells the greater production of the transmitter, acetylcholine, when they are active than when they are not."

These are remarkable achievements indeed, and only a clown would call the manner of achieving them unsophisticated. But widespread application of the results can come about only after scientists have learned to deliver the desired chemicals to the desired sites more easily. They are prevented from doing so by a set of circumstances which exist nowhere else in the body—the so-called "blood/brain barrier."

Though the brain constitutes only one-fiftieth of the body's total weight, it requires a full one-fifth of the body's oxygen-rich blood supply. If circulation is cut off for more than a very few minutes, the result is permanent brain damage and, in a few more minutes, death. Since the brain's cells are constantly bathed in blood, it would seem that the simplest way to get drugs to the brain, just as to other sites in the body, would be to put them into the bloodstream, either orally or by injection.

214

But the brain's cells are uniquely surrounded by a little-understood electrochemical fence, the blood/brain barrier, which admits only certain selected substances and keeps out everything else. Until this barrier is overcome, the cells will simply not take from the bloodstream many of the chemicals which, when injected directly into the cells, have such profound effects.

Even so, the blood/brain barrier does permit the passage of a varied inventory of substances, and it is no news that some of these substances can influence the mind's thoughts and perceptions, and hence the person's behavior, in striking ways. The Chinese described marijuana and its effects as early as the twenty-eighth century B.C. Over the centuries of recorded history there has hardly been a time or a place without some knowledge and practice of opiates or stimulants of one kind or another. Even the poorest people, even in the most primitive societies, have known where in the plant kingdom to seek solace or a moment of borrowed ecstasy. From the poppy fields of the Near and Far East have come opium and its derivative narcotics, morphine and heroin. From the female hemp plant, *Cannabis sativa,* which will grow anywhere in the temperate zones of the world including backyards and window boxes, come hashish and marijuana by all their multifarious names. Nutmegs and morning-glory seeds, cacti and coca leaves, have consistently provided kicks and calms for the inhabitants of the regions where they grow.

The news is that in recent years scientists have been raiding the herbals of folk medicine, testing the efficacy of many ancient drugs, finding new uses for them, extracting the active elements from the grosser content, synthesizing the vital chemicals, and creating new, wholly synthetic drugs in the laboratory. They have been getting down to the basic biochemistry of these substances as well as the brain's own key chemicals, and studying their complex interactions. They have built up an

215

impressive arsenal of psychochemicals or "mind drugs"* and given birth to the lustily growing new science of psychopharmacology.

While these developments proceed, plenty of mind-affecting drugs are already on the market, and many that aren't, are on the black market. Dr. Donald Louria, Chairman of the New York State Council on Drug Addiction, estimates that some nine to thirteen billion sedatives, tranquilizers, and stimulants were manufactured in the United States in 1965. "This," he calculates, "means 35 to 60 pills or capsules for every man, woman, and child!" These are legitimate prescription drugs, though many find their way into illegitimate channels where they are dangerously misused and overused. There are, too, the frankly illegitimate drugs, most notoriously the narcotics and especially heroin, that claim their annual toll of misery, bondage, and death through addiction and overdosage.

Finally, there are the hallucinogens, also known as psychedelics—principally marijuana, which is relatively mild in its effects, and LSD, which is explosively potent. These are, at the moment, the most controversial of drugs because they have gained wide popularity among young people.

When doctors talk about drug abuse, they are usually referring to drugs that are self-administered, taken through the user's own volition. But, as in the case of ESB, there is considerable concern among scientists about the potential abuse of future psychochemicals in terms of the powers they might give the clever and ruthless over their fellows. Part of this concern is due to a familiarity with certain aspects of psychochemistry

---

* Most people seem to know that the effects of these drugs are different in different people, but Dr. Elkes believes it cannot be emphasized too strongly that even "the same drug, in the same dose, in the same person may produce very different effects, according to the events which precede or follow a particular medication."

being investigated by the world's military establishments— whose fascination with the mind drugs is hardly less than that of the psychopharmacologists themselves, though for different reasons. Major General Marshall Stubbs, at the time chief of the Army's Chemical Corps, told a congressional committee: "The characteristics we are looking for are . . . exactly opposite to what the pharmaceutical firms want in drugs—that is, the undesirable side effects."

Most of what interests the military is highly classified. However, Robert Coughlan in *Life* was able to offer an imposing catalogue of "incapacitating agents" being actively pursued.

> The things these agents can do now are many and exceedingly strange. Besides the hallucinogens there are, for instance, *euphoriants* [italics mine]. They incapacitate by making their victim so witlessly optimistic about everything that he is no good for anything. As one Army medical attendant at a Chemical Corps tryout on human volunteers explained, "Even the worst food, like Army food, tastes absolutely delicious to them. They'll tell you it is the best they have ever eaten!" The opposite number of the euphoriants is the *depressants*. These drugs induce morbid gloom and prevent the victim from doing anything because he feels that nothing is worth doing. Also there are *cataplexogenics*. Their victim remains fully conscious, thinks normally and tries to respond to stimuli in his usual way, but he finds that his muscles don't obey. They might be rigid, flaccid, or limp, but in any case they are useless and he is immobilized.
>
> Then there are the *disinhibitors*. These block or weaken the controls that normally keep behavior on a fairly even keel; the victim overreacts, with excesses of talking, imagination, emotions, and actions. (Alcohol is a familiar disinhibitor but a relatively mild one). In addition there are the *chronoleptogenics*. They distort the sense of time and since the victim cannot discriminate between hours and seconds, he loses track of relationships in which time is involved, becoming ineffectual and lost. And there are the *confusants*. They cause the victim to lose track of all relationships; the

217

world is totally out of joint and everything in it (himself included) is uncertain, contradictory, overwhelmingly strange, and perplexing.

These descriptions have been generalized, of course. In practice the effects are variable and subject to many limitations. None of the compounds is 100 percent effective in all circumstances, and some are only moderately effective even under the best circumstances. The important point about all of them, however, is that they do exist, they do affect the brain and they do manipulate specific aspects of behavior. The rest is a problem of product development, so to speak— of tinkering, refining, and improving, adding a new twist here and there to make them better and better—or, perhaps one ought to say, worse and worse.*

Among yet other possibilities, Coughlan makes ominous mention of "chemicals that increase suggestibility and hence could be extremely useful in 'brainwashing' prisoners of war or even (if diffused in water supplies, or perhaps in common table salt, as is done with iodine) in making whole populations receptive to propaganda. The Chemical Corps, through its liaison program with industry, receives hundreds of odd compounds monthly for testing and there is no telling what will turn up."

All of these new means of tampering with the human brain and behavior via electricity and chemistry invoke the same kind of fears and the same kind of moral dilemmas as those aroused by the biomedical developments explored in Parts I and II. They involve our definition of the good life, the role of the individual, the assignment of power and authority—and their limitations and restrictions.

We all have qualms about those in power inflicting their will

---

* In the movie *Goldfinger,* it was some such incapacitating agent, all ready and perfected for cinema purposes, which the conspirators arranged to have sprayed on Fort Knox to put the guards out of effective action.

on the unconsenting masses. But what of those who choose to exercise the new controls on themselves, and who insist on their individual right to do so? Suppose a man wants to have electrodes implanted in the sexual centers of his brain and carry a self-stimulating pushbutton device to turn on his desire and capacity whenever he pleases—should this be denied him? It is certainly not unusual for a patient suffering from feelings of sexual inadequacy or impotence to go to a urologist or psychotherapist for help; and the doctors do try hard to find remedies. If a physician decided that the implantation of electrodes was the easiest remedy—and quite safe—for a given patient, it might appear to be merely an extension of normal medical advice to send the man to a neurosurgeon and have it done. But many doctors would have great ethical qualms about proceeding.

In the controversy over the use and abuse of psychedelic drugs many intellectuals and artists have insisted that marijuana and LSD have many positive values to recommend them: insights into the self, expanded awareness, enhanced creative potential. Many argue for the legalization of these drugs on the ground that they are not addictive narcotics, and that their troublemaking potential is certainly no greater than other universally accepted commodities such as caffeine, alcohol, and nicotine.

One of the least investigated areas of the drug problem in our society is the "white-collar drug scene," a label coined by Bruce Jackson in *Atlantic*. Jackson gives a quietly hair-raising account of a pill party he was invited to in a major American city. The host, on pep pills, had not slept in three or four nights. Instead of the usual cocktail-party bar, there were candy dishes full of many-colored pills and capsules, most of them amphetamines and barbiturates available by prescription. Also on hand was a well-worn copy of the *Physicians' Desk Reference,* a handbook on pharmaceuticals for doctors. The pills were passed around, and the guests selected from them as they would from a tray of assorted cookies. It was not a wild party. There was music, but no drunkenness or sex play. None of the partygoers were teen-agers, college undergraduates, or hippies. They were all well-dressed, well-behaved, well-educated, and presumably mature adults in the middle-income bracket, some of them married couples who just happened to become "pillheads." They went regularly to such parties where the conversation centered mainly on the varieties of drugs, how to enhance their effects, how to insure a continuing source of supply to feed their habits.

Most of these people seemed to have no clearcut idea as to why they had become habituated to the pills, except that they consider life without the pills either too complex to cope with or too boring to tolerate. The amphetamines and barbiturates they use are all substances which get through the blood/brain barrier to influence the chemistry of the cells in those primitive areas of the brain that Miller and Elkes were talking about. The white-collar pillheads also use hallucinogens occasionally, smoking marijuana and taking LSD, substances that get through to some of the brain's higher centers as well, distorting and heightening perceptions, thoughts and feelings, sometimes intensely and for prolonged periods of time.

"Lately," writes Jackson, "attention has been focused on drug abuse and experimentation among college students. Yet all the college students and all the junkies account for only a small portion of American drug abuse. The adults, the respectable grown-ups, the nice people who cannot or will not make it without depending on a variety of drugs, present a far more serious problem. For them the drug experience threatens to disrupt or even destroy life patterns and human relationships that required many years to establish.

"And the problem is not a minor one," he warns. "Worse, it seems to be accelerating." One of his pillhead friends told him one night, "You better research the hell out of it because I'm convinced that the next ruling generation is going to be all pillheads. I'm convinced of it. If they haven't dysfunctioned completely to the point where they can't stand for office. It's getting to be unbelievable. I've never seen such a transformation in just four or five years. . . ."

As time goes on, and as the biochemists and psychopharmacologists pursue their research, the pillheads will have at hand an increasingly sophisticated array of psychochemicals to draw from. The noted psychologist, Dr. B. F. Skinner of Harvard, predicts that "in the not too distant future the motivational and emotional conditions of normal daily life will probably be maintained in any desired state through the use of drugs." Is this a good or bad thing? Aldous Huxley, in his younger days—long before the prospect seemed to have any basis in reality—bitingly satirized the whole notion in *Brave New World*. But he lived to wax lyrical in its favor in *Doors of Perception,* the result of his experiences with mescaline.

Do individuals have the inalienable right, as many argue, to take any drugs they please whenever they please—especially if the drugs are nonaddictive—without interference from the law, or, for that matter, from their physicians? Should free individuals not be the sole guardians and custodians of their own inner experiences? If a man chooses to sit in a room and quietly enjoy his drug-induced visions, or whatever stimulation or lethargy the drug of his choice brings him, is it anyone else's business?

In the present state of psychopharmacology, yes, it is other people's business. Taking these drugs in an unsupervised milieu, in large dosages, and on a continued basis, is more than a little risky. The taker may do himself irreparable damage physically, psychically, and socially. He may be unable to function or handle his responsibilities, and not care one way or the other whether he does or not, thus leaving society to worry

221

about him and his dependents. The risk is not all his own, because the inner experience he chooses to undergo can have devastating consequences for others, including his wife and children. Moreover, under the influence of drugs he may very well be a genuine menace to innocent bystanders. Drugs may distort his perceptions so that he is unable, say, to drive a car properly; yet his judgment may also be distorted so that he believes he is handling the car even more expertly than usual. He may feel a soaring sense of power, a delusion that can make him reckless and bring injury or death to himself and others. Or the drug may bring out latent paranoid tendencies, and, fearing an imaginary attack, he may attack first. So drug-taking cannot be a person's purely private affair.

Nevertheless, questions about the internal freedoms of the individual are valid enough. We do trust people to drink whiskey, which is a mind-affecting drug—and we hold them responsible for the consequences, such as drunken driving. Why not trust them, in the same way, to smoke marijuana? Such questions will be even more valid, and more plentiful, in the years ahead. Drugs will presumably become more selective in their action, producing the desired moods and perceptions without damaging the user or curtailing his ability to function normally. (Our concepts of what constitutes "normal" functioning are also due for some changes, of course.) When these things come to pass, we may be hard put to attach any sense of moral wrongdoing to the mere taking of drugs.

Is it not, after all, one of medical science's main purposes to provide us with medications to make us feel better? No one thinks it is wrong to eat whichever foods will most nourish our cells and coax them to their maximum metabolic efficiency. Yet foods are nothing but chemical substances derived from plants or animals that we grow or slaughter, and which we take in our bodies for the purpose of doing us good. In recent years more and more of our foods and our food supplements have become partially synthetic, which does not seem to have rendered them unacceptable. Why should it be considered unnatural, then, to take in any chemical substances that will do us good, especially if the side effects are negligible, even if they happen to be labeled as drugs instead of as food? Even foods are not devoid of side effects. They can, for instance, cause nausea and stomachaches. They can line our arteries with cholesterol. They carry small quantities of pesticides and radioactivity into our system. When we were children, we were all, at one time or another, encouraged to eat foods that we were told (probably erroneously) were good for our brains. Well, that's what the psychopharmacologists are working on.

A real danger inherent in promiscuous drug-taking, of course, is that people might become so enchanted with their drugged states that they have little desire for experience in the real world—a world which does not interest them, perhaps, because it seems both dull and hopeless, a world they were eager in any case to retreat from.

It is too bad that a world so full of intrinsic fascination and adventure can seem so hopeless and uninteresting to people who are neither ill nor poverty-stricken. Perhaps our creative people in all fields of endeavor, from politics to the arts, will want to exert themselves a bit to see that society begins to make better sense again, and that hope begins to seem worth hoping for again. What are people for? What are our human goals and values? What ought they to be? The questions repeat themselves. When we come up with answers that begin to satisfy us, perhaps we can then start building a society whose members will have little need and less desire to retreat from it via the drug route.

Even if individuals could be counted on to use drugs sanely and judiciously, this would only take care of the hazards of self-administration. How we use drugs on ourselves is one thing; how they are used on us is another. The mere existence of a versatile armamentarium of psychochemicals that can bring pain or pleasure, sleep or wakefulness, sexual desire or impotence, feelings of heat or cold, thirst or hunger or satiety; that can offer greater insights and intellectual powers as easily as they can deaden or disorient the mind; that can, in brief, control human brains and therefore human behavior in almost any desired way, holds out prospects that are not guaranteed to cheer us. Whose hands will they fall into? How can we insure that they will be used for our benefit, and not for the selfish or criminal purposes of private parties or of nations? The late Lord Brain voiced his hope "that the scientific freedom which produces this knowledge will act as an effective antidote for its misuse," but he admitted, in the same sentence, that "our experience of nuclear weapons may justify some skepticism about this."

If scientists who look to the future seem to dwell on the hazards they foresee, it is only, as Dr. Lederberg explains, because "we must try to anticipate the worst anomalies of biological powers. To anticipate them in good time is the first element of hope in developing institutional and technological antidotes."

Obviously good things as well as evil can and do ensue from scientific advance. As just one example, a real breakthrough in understanding the specific molecular workings of the body's immunological systems would speed the conquest of disease all the way from the common cold to cancer, would overcome the transplant barrier, and would provide key insights into the nature of allergy and the process of aging. This kind of breakthrough in understanding could come at any time, and if we ourselves were not the immediate beneficiaries, our children certainly would be.

Those optimistic scientists who believe that we will not necessarily be annihilated in a nuclear war nor succumb to the poisons and pressures of a polluted and overpopulated planet offer great hope that human beings of the future will be much improved in body, mind, and spirit, better able to enjoy their joys and better equipped to overcome their troubles.

Brain research, apart from the dangers so far emphasized, also promises many good things. These good things, as is the case with the results of most scientific research, are of two kinds: One is the elimination of defects and diseases; the other is a positive upgrading of the normal.

The diseases and defects of the mind have always seemed to be of a fundamentally different nature than the diseases and defects of the body. A man harassed by the occurrence of mental illness in his family might cry out in great anguish for some easy remedy. But chemistry never held out much promise of the prayed-for solutions. When Macbeth inquired after his wife's health, he knew the question was purely rhetorical, bound to elicit a useless reply:

MACBETH: How does your patient, doctor?

DOCTOR: Not so sick, my lord,
As she is troubled with thick-coming fancies
That keep her from her rest.

MACBETH: Cure her of that:
Canst thou not minister to a mind diseased;
Pluck from the memory a rooted sorrow;
Raze out the written troubles of the brain:
And with some sweet oblivious antidote
Cleanse the stuft bosom of that perilous stuff
Which weighs upon the heart?

DOCTOR: Therein the patient
Must minister to himself.

MACBETH: (*While putting on armor to go meet the English*)
. . . If thou couldst, doctor, cast
The water of my land, find her disease,

225

> And purge it to a sound and pristine health,
> I would applaud thee to the very echo,
> That should applaud again.

We are luckier than Macbeth. Today, doctors oriented toward psychopharmacology can provide much better answers than his physician could. Since psychopharmacology is barely in its infancy, future doctors should be able to do much better by us.

In the latter part of the fifteenth century, around the time Columbus was negotiating with the rulers of Spain to finance his quest for a westward route to the Indies, two scholarly monks were writing a book of horrors called *Malleus Maleficarum*. Its publication gave sanction to the notion that anyone afflicted with what we now call mental illness was possessed by demons and witches. As a result, uncounted hundreds of thousands of innocent wretches whose ailments, if any, were diagnosed or misdiagnosed as being in this category, were subjected by the pious to unspeakable cruelties. This went on for centuries, with only an occasional voice speaking out, at no small risk to the speaker, in ineffectual protest. As late as the eighteenth century, when a humane doctor named Philippe Pinel arrived at La Bicêtre, the famous French mental hospital, he created a sensation by taking the vermin-infested patients out of their chains.

In our enlightened era the snake pit has given way to the analyst's couch and other therapies, and psychosis has been recognized as illness rather than demonic possession. But until recently it was still largely regarded as a special, separate form of illness. It was "functional"—the result of the patient's environment, his relationships with other people, the circumstances of his upbringing and experience. Since it was "mental" rather than organic illness, it could only be reached, if it could

226

be reached at all, by somehow reaching the disturbed mind. The idea that chemistry could provide any *substantial* answers would probably have been scoffed at by most Freudian psychiatrists, though not by Freud himself, who once wrote: "The future may teach us how to exercise a direct influence, by means of particular chemical substances, upon the amounts of energy and their distribution in the apparatus of the mind."

In the early 1930's a young Viennese psychiatrist, Dr. Manfred Sakel, noticing that an accidental overdose of insulin cleared up the mental disturbance of a diabetic drug addict, decided to try insulin shock therapy on his schizophrenic patients. Some of them improved dramatically. The treatment was controversial, and no one could explain how it worked. But insulin therapy is still used in many places today, and is regarded by many psychiatrists as the milestone that proved mental illness could be helped by purely biological treatment.

Actually, such evidence was already at hand much earlier in the century, but the evidence was largely unrecognized as such. The evidence I am talking about was the discovery of the syphilis spirochete in the brain tissues of victims of a widespread form of insanity known as paresis. "The young residents who are now entering the field of psychiatry," says Dr. Percival Bailey, Director of Research at the Illinois Psychiatric Institute, "cannot imagine the rows of paretic incontinent odorous Caesars who filled the wards of the mental hospitals fifty years ago. Many and strange were the theories which attempted to explain their plight, ranging from demonical possession to excessive venery, until Bayle demonstrated the leptomeningitis which accompanied the disorder, and Moore and Noguchi found the spirochetes in their cerebral cortices." Thus paresis, one of the severest and commonest types of mental illness in its day, was shown to be caused by the ravages of an infectious microorganism on the victim's nervous system. When substances like arsphenamine or penicillin were used to treat or prevent syphilis, they were also being used to treat or prevent paresis. Here was a drug therapy for a mental disease, but no one at the time followed up the idea as applicable to other forms of mental disease. Now that such studies are being vigorously pursued in the attempt to track down the causes of schizophrenic reactions, Dr. Bailey predicts that "whoever finds the key to them will empty our mental hospitals as happened when syphilis was conquered."*

Another ailment that should have pointed the way sooner toward psychopharmacology was pellagra, a disease once quite common in lands bordering the Mediterranean, as well as among undernourished inhabitants of some areas in the American South and among chronic alcoholics everywhere. One of the characteristic symptoms of pellagra was a serious mental derangement. Scientists were a long time finding the cause of pellagra. Dr. Gerard likes to cite those who early found "evidence" that pellagra was hereditary because it ran in families. But the reason it ran in families was that all the members of the family ate the same poor diet. The cause turned out to be lack of a crucial nutrient—niacin or nicotinic acid, one of the B-complex vitamins. When this vitamin was supplied, the pellagra was cured, and all the symptoms, including the mental derangement, disappeared.

More recently a third type of mental illness was traced to purely physiological causes. It is the form of mental retardation known as phenylketonuria (PKU), first recognized as a distinct disease by a Norwegian scientist who noticed a peculiar

---

* True, syphilis has been "conquered" in that medicine now knows the cause of it, and some effective remedies for it. But the disease is far from eliminated; in many places, and especially among young people, it is on the rise—partly because the spirochetes have been developing resistance to the antibiotics, and partly because people, *believing* that syphilis has been conquered, have grown careless.

smell about the children afflicted with it. The protein metabolism of these children was faulty, it was later discovered. Because of a genetic defect, they lack a liver enzyme which catalyzes the conversion of one amino acid (phenylalanine) into another (tyrosine). The missing enzyme doomed them to retardation and early death. Their bodies and brains could not handle the excess of unprocessed phenylalanine, and they were, in effect, poisoned. Now there is a test that can detect the condition at birth. By keeping phenylalanine out of the child's diet during the crucial ages, PKU can be warded off, and the child's development is virtually normal. There now exist families where one child is retarded and the other is not, simply because knowledge of the causes of PKU came too late for the one but not for the other.

Here, then, are three "mental" diseases—paresis, pellagra, and phenylketonuria—one caused by an infectious microbe, one by a vitamin deficiency, one by the hereditary lack of a key enzyme; ample proof that at least some mental disorders can be caused by biology, and are amenable to treatment by biology.

What really brought psychopharmacology into thriving prominence, though, was the development of tranquilizers. The first of these was reserpine, isolated in 1952 from the dried roots of *Rauwolfia serpentinum,* known and used in India for the same purpose as early as 1000 B.C. Fast on the heels of reserpine came chlorpromazine, meprobamate, and other popular varieties. "By reducing the hostilities and rages of patients, and by quieting their fears and anxieties," writes Isaac Asimov, tranquilizers "reduce the necessity for drastic physical restraints, make it easier for psychiatrists to establish contact with patients, and increase the patients' chances of release from the hospital." Because of these characteristics, their impact on psychiatry was immediate.

"Within a year after the start of large-scale application, the atmosphere of the mental hospitals was radically improved," said Dr. Henry Brill, Deputy Commissioner of Mental Hygiene for the State of New York, at a 1962 symposium. "Disturbed behavior was no longer a problem. Within a few years, restraint had been 90 percent reduced and by the end of the first year the traditional population increase of 2,000 was replaced by a decrease of 500."

Soon after the tranquilizers came another class of drugs, the antidepressants, also known as "psychic energizers" and "mood elevators," the first of them pioneered by Dr. Nathan S. Kline of New York's Rockland State Hospital.

In the ancient worlds of Greece and Rome, mental troubles were usually put in two categories: mania and melancholia. The new drugs supply a remedy for each—tranquilizers to quiet the manic, antidepressants to lift the melancholic. As a result of the combined effects of tranquilizers and antidepressants, says Dr. Brill, "by 1959, 60 percent of all New York state mental hospital cases were on biological treatment . . . This time the public health impact was clearcut and significant . . . The story was repeated throughout the Western world with some variations."

Much excitement was raised, too, by studies of mescaline, LSD, and other hallucinogenic drugs. Because they evoked symptoms that seemed to mimic those of psychotic states, they were called psychotomimetics. It was hoped that a study of these "artificial psychoses" could give clues to the cause and cure of true psychoses.

Meanwhile psychopharmacologists were off in hot pursuit of biological approaches to the most common and troublesome of today's serious mental ailments, schizophrenia. The Greek-derived word means split-mindedness. It was coined by the Swiss psychiatrist, Dr. Paul Eugen Bleuler in 1911* because the

---

* The older name for it, *dementia praecox,* (dementia that comes on early) was coined by Dr. Emil Kraepelin in 1898, to distinguish it from the dementia that comes with senile degeneration.

disease was characterized by a splitting off of at least a part of the patient's mind from the rest of his personality, a withdrawal into a delusion-filled world of its own, separate from the real world in which others around him can participate. More than half the present population of mental hospitals are schizophrenics, and the disease seems to be common throughout the world. No class, no race, no nation, no geographical region seems to be immune. The actual manifestations of schizophrenia are so varied* that it is no longer considered to be a single disease. Psychiatrists often speak of "the schizophrenias" as a class of related diseases.

More and more researchers are becoming convinced that at least some forms of schizophrenia are rooted in biology, and are therefore treatable, and perhaps curable, via the biological approach. There is certainly general agreement that chemical therapy can at least be a valuable adjunct to other forms of treatment. It is believed, too, that at least a *tendency* toward schizophrenia can be inherited**—the tendency being some sort of metabolic weakness that permits malfunctioning of the brain's enzymatic apparatus, especially when the victim is under intolerable stress.

Dr. Linus Pauling was speaking for more scientists than himself when he said, in 1962, "I believe that most mental diseases are molecular diseases, the result of a biochemical abnormality in the human body. I think that the mind is a

---

* Three of the major varieties, named by Kraepelin and still considered valid, are *hebephrenia* (childmindedness), *catatonia* (withdrawal into a stolid silence), and *paranoia* (a compound of suspicion and hostility, with delusions of persecution).

** About one percent of the population is afflicted with schizophrenia (which means there are at least thirty million schizophrenics in the world). So chances of coming down with it are 1 in 100. But if your brother or sister has it, your chances go up to 1 in 7. If your identical twin has it, they go up to 3 in 4. These figures cannot be regarded as proof positive, however. Remember that some scientists used to think pellagra was hereditary because it ran in families.

manifestation of the structure of the brain, that it is an electrical oscillation in the brain, supported by the material structure of the brain; I think that the mind can be made abnormal by an abnormality in the chemical structure of the brain itself, usually hereditary in character, but sometimes caused by an abnormality in the environment."

Metabolism and enzymes have been mentioned before in this book, but now a fuller explanation of metabolism and the role of enzymes seems in order as a prerequisite for appreciating the direction of current research in psychopharmocology. Normal metabolism requires the carefully regulated balance of innumerable processes involving the constant buildup, breakdown, and rearrangement of multitudes of molecules according to the body's needs at any given moment. To achieve this monumental harmony, myriads of chemical reactions must take place simultaneously at trillions of locations in the body—all delicately orchestrated in a complexity beyond description or imagination. Every one of these multifarious reactions is dependent upon enzymes—which are thus the key to all life.*

Enzymes are large, highly specialized protein molecules. They do not participate in biochemical reactions themselves. They act rather as catalysts. They lend energy to speed the actions along. In the human body a given enzyme—and, as a rule, only that enzyme—can perform a given catalytic action, and only that single action. If the enzyme fails to carry out its task (or is missing), the whole chemical process comes to a halt. The seriousness of the consequences depends on the importance of the process. If it is a vital one (as in the case of PKU), the person might sicken or die.

---

* The virus, on the borderline of life, is kept from crossing the border by the lack of any mechanism for making enzymes. Only when the virus sneaks into a living cell and takes over the cell's enzyme-manufacturing apparatus does it spring into a gaudy, illicit life of its own.

Enzymes usually work in teams, each performing a single step of a given chemical reaction. Each enzyme does only a little bit: It may help remove from, or add to, a given molecule no more than two or three atoms before another enzyme goes to work on it. To maintain the intricate metabolic counterpoint of buildup and breakdown, whole populations of enzymes keep busy around the clock. In the liver—the biochemical factory where so much of our food is broken down and converted into substances the cells can handle before it is released into the bloodstream*—a single cell may contain more than a thousand different kinds of enzymes, and there may be thousands of each kind.

The reasons so many enzymes must be kept employed carrying out chemical reactions at fantastic speeds but in tiny little steps is that, if substances were broken down all at once, the result would be an explosive release of energy, and the blood would boil in a burst of spontaneous combustion. For life to go on, the body must burn its fuel (food) at a slow, measured pace. If the burning were not kept in check, the body's fuel would act like the uranium in an A-bomb. Thanks to enzymes, it acts more like the uranium in a controlled nuclear reactor.

The precise and flawless functioning of enzymes, then, is crucial to the body's (and therefore the brain's) welfare. Each enzyme must watch for its cue, then move in unerringly and without hesitation to fulfill its role. Nature seems to have designed each enzyme molecule—an incredibly intricate three-dimensional structure—so it would exactly fit, just as a key fits a lock, the one-and-only molecule it is meant to work on, at precisely the moment it is meant to work on it.

Without a clearcut, specific duty to perform, an enzyme could be dangerously versatile. If it had to choose instanta-

---

* There is a persistent history of belief that liver function is somehow bound up with mental disease, and some researchers are again becoming intrigued by this idea.

neously between several possible alternatives, it might choose the wrong one—or the right one at the wrong time. That such confusion is possible has been proven time and time again in the laboratory. As one classic example, there is an enzyme called succinic acid dehydrogenase whose sole task is to catalyze the removal of a pair of hydrogen atoms from the succinic acid. This is part of a chemical cycle that takes place in every cell. But if, in a test tube, you add malonic acid as well as succinic acid, you will have set the stage for subversion. The malonic acid molecule is quite similar in structure to succinic acid, and both substances are now competing for the attention of the same enzyme. The dehydrogenase can be fooled. It often goes to the malonic acid by mistake. When it does, it is no longer available to the succinic acid, which then goes around unable to get rid of its excess hydrogen atoms. If this were to happen inside a cell, a vital cycle would come to a halt.

Scientists believe this sort of thing does occasionally happen in the body, and in the brain. The competing substance might be a poison or a drug, for example. It lures the enzyme to itself, thus inhibiting its normal activity. Such a substance is called an *antimetabolite,* since it interferes with metabolism, and the process—a key concept in the current assault on mental disease—is known as *competitive inhibition.* The antimetabolite competes for the enzyme and, winning the competition, inhibits the enzyme's normal function. Many of the mind drugs mentioned earlier seem to be antimetabolites, achieving their bizarre effects by competitive inhibition. In the case of a drug administered from the outside, such as mescaline or LSD, the effects usually wear off when the drug does. But if the body is manufacturing its own antimetabolites—why, they're there all the time, and the person is said to have a disease.

These ideas, as applied to mental illness, are not brand new. Even before the turn of the century, Dr. Louis Lewin hinted at a significant tie-in between the effects of certain drugs and the symptoms of mental illness. Kraepelin suggested that dementia praecox might be due to "an autointoxication in consequence of a disorder of metabolism."* And Dr. Carl Gustav Jung postulated a mysterious "toxin-X" that he thought might be at the physiological root of mental troubles. But all these theorizings occurred at a time when they could not really be fruitfully pursued—back before the flowering of biochemistry, and long before the outbreak (one might almost call it that) of molecular biology.

Current investigations in psychochemistry show that there are surprisingly few substances involved in the brain's vital functioning. A couple of the important ones are adrenalin and serotonin. In the interplay of thought, mood, and perception, these substances are being constantly built up and broken down by enzymes. An antimetabolite that could effectively engage and competitively inhibit any of these enzymes could halt the buildup or breakdown of adrenalin or serotonin. With either too much or too little of these substances in key areas of the brain, major psychic disturbances might be expected to ensue. Since these disturbances would be based on physiological changes, it follows that the chemistry of, say, schizophrenics should be abnormal in some ways. And so it turns out to be.

The serum of schizophrenic patients, and substances isolated from their blood and urine—substances apparently absent in normal people—have brought on quite disturbing physiological and behavioral effects in a variety of animals, from tadpoles to white mice to monkeys, and have even interfered with cell division in plants. Though all this was tantalizing, enthusiasm

---

* It may well be that mystics, after mortification of the flesh through fasting and other bodily deprivations, see their visions because their cerebral metabolism has been disturbed in a similar manner. Dr. J. D. Bernal once wrote: "Altering in any perfectly sound physiological or surgical way the functioning of the body will certainly have secondary but far-reaching effects on the mind."

has been cautious because no one could pin down the toxin-X, if any, that was responsible for the disorders, the effects were inconsistent, and different researchers got conflicting results. It was all so confusing, in fact, that Dr. Manfred Bleuler wrote: "It is possible that as a consequence of these negative results the search for a specific somatic basis for schizophrenia will be given up for a long time to come, if not permanently."

The irony is that Dr. Bleuler wrote those words in 1951, just when a new and concerted effort was getting under way to find that "specific somatic basis for schizophrenia." Notably in Canada, three members of the Saskatchewan Committee for Schizophrenia Research—Dr. Humphry Osmond, Dr. John Smythies, and Dr. Abraham Hoffer—initiated research that considerably illuminated the relationship between drug-induced states and true mental disorders. They also showed that substances producible in a test tube—such as adreno-chrome, which strongly resembles adrenalin—can fool the enzyme, amine oxidase, and successfully engage it so that it is no longer free to perform its main function—breaking down adrenalin—the result being an excess of adrenalin.

In the middle and late 1950's, Dr. Heath and his associates at Tulane succeeded in isolating a material they called *taraxein* from the blood serum of schizophrenics. When injected into normal volunteers and into former schizophrenics whose symptoms had disappeared, taraxein induced schizophrenic symptoms artificially. Here is Dr. Heath's own description of what happened:

> All subjects developed symptoms which have been described as being present in schizophrenia. Consistently, primary or fundamental symptoms appeared. Blocking and thought deprivation developed; all subjects were autistic and complained of depersonalization. They appeared dazed, with diminished contact to the environment; had a blank look in their eyes; and showed a lessening of animation in facial expression. These primary or fundamental symptoms appeared

first. They were the most consistent and developed even
with the administration of taraxein which had given minimal
activity by animal assay. Every classical secondary symptom
was observed in our group of test subjects. Their symptoms
included catatonic stupor and excitement, hebephrenia, ideas
of reference, delusions of persecution, grandiosity and auditory hallucinations.*

It would appear, then, that some chemical or complex of
chemicals present in the blood of schizophrenics is powerful
enough to induce temporary schizophrenia when injected into
the bloodstream of normal people. But the applause, if any,
was muted. "It is indicative of the climate of opinion in our
day," commented Dr. Osmond later, "that this was welcomed
with indignation and incredulity rather than with joy." Others
have corroborated the Tulane results, but taraxein still remains
controversial—and will probably continue to be until its exact
chemical composition can be determined. When this is done,
and when the substance is synthesized, and if it indeed turns
out to be the Toxin-X that causes schizophrenia, and if an
antidote can be devised, then that would be the end of schizophrenia.

All these ifs, of course, are iffy indeed. To begin with, because schizophrenia is believed to be a series of illnesses rather
than a single entity, there may be multiple causes. And even if
it turns out that all schizophrenia has some basis in physiology,

---

* Dr. Heath has also been able to induce psychotic symptoms by administering other chemicals, and by ESB. After seventeen years of research at Tulane, he summed up his thinking in a 1966 paper:
". . . schizophrenia has been postulated to represent a biologic irregu-
larity that predisposes to inheritance of the disease. According to this
hypothesis, a biochemical abnormality, created by the error, influences
cellular function in precise regions of the brain and produces a physiologic abnormality which underlies the behavioral disorder. Our observation suggest that the biologic abnormality so alters the capacity to handle
stress as to intensify the secondary behavioral symptoms."

there is, as Gilbert Cant of *Time* has pointed out, a big chicken-or-egg question here. Which came first? Illnesses can be somatic, psychic, or psychosomatic in origin. Doctors know, for example, that an ailment like asthma can be brought on by emotional problems; but then the asthma causes real physical damage that cannot necessarily be made to disappear when the emotional problems are solved; in fact, the physical ravages of the disease are likely to cause further emotional disturbances. That is the way it could be with schizophrenia. The derangement might easily be caused by purely psychic or environmental factors—which in turn cause physiological changes, which may in turn perpetuate the derangement. All this needs to be worked out. It seems likely that schizophrenics will continue to need psychiatry as well as medication in a combined approach. But the new understanding of the biology underlying mental illness now makes it as feasible to think of eliminating psychoses as to eliminate, say, cancer. Apart from the alleviation of human suffering, which nearly all of us would count an unalloyed blessing, the savings would be incalculable—in terms of the money it costs to maintain people in mental institutions, in terms of useful man-hours lost to society, in terms of emotional erosion of families—even the disruption of nations. After all, had there been a drug to counteract psychotic hostility and aggression (as there may well soon be), and if it had been given to Lee Oswald, might John F. Kennedy still be alive? And had it been possible to treat Adolf Hitler, how might the course of twentieth-century history have been altered?

Such a drug may already exist—though it remains to be proven how widespread its effective application might be—in diphenylhydantoin (DPH). This drug has been on the market for some thirty years as a treatment for epilepsy, but its possible uses as a specific remedy for electrical disorders that may cause anger, fear, depression, anxiety, and a kind of chaotic nonstop compulsive thinking, was only recently brought to light through the interest and determination of the financier, Jack Dreyfus, Jr.—who has set up a research foundation to pursue more intensively the nonepileptic benefits of DPH.

If DPH, or any similar or superior drug, were indeed a specific for fear and anger, might it be administered to keep angry people from rioting in the streets? Would that be a good idea—or not such a good idea? As an interesting aside to all this, one of the remarkable observations recently made about drug effects is that they differ not only among individuals, and not only in the same individual at different times—but they differ considerably when the individual is in a group. In animal studies, for instance, it was found that to induce a certain level of excitement required a certain quantity of a given chemical—when the animal was alone. But when the same animal was part of a group, especially an active group, a much smaller quantity of the chemical was required to induce the same level of excitement. This might tell us quite a bit about mob behavior in human beings.

The ability to turn off mental troubles was only one of the twin benefits promised earlier as likely to come out of what we might call electropsychochemistry. The other promise was a positive enhancement of normal functioning—heightened moods and perceptions, more creative imagination, greater intellectual capacities.

To a small extent, some of these possibilities have been realized. Normal people who have experimented, under supervision, with the mood-elevating and psychic-energizing drugs have claimed increased powers of concentration. Many say that, while under the drug's influence, they not only feel better but think better and learn more easily. The hallucinogenic or psychedelic drugs have produced even more eloquent testimonials, some of them flamboyantly lyrical in their enthusiasm. Witness Aldous Huxley's heartfelt endorsement of the mescaline experience. Witness also the many evangelists for LSD, who testify that they have achieved profound insights into themselves, have sharpened their sensibilities, have learned better to appreciate their fellow man, have become better all-around people—these good effects lasting not merely as long as the LSD "trip," but for a long time thereafter, perhaps for a lifetime.

As of this writing, because of public furor over the abuse of LSD, bona fide research may be slowed down—but probably not for long. Not that LSD's potential harmful effects on the psyche of the careless user—or the more recently discovered evidence of possible damage to the chromosomes—have been exaggerated; they have not. But its potential rewards are too great to be ignored. We certainly cannot dismiss all this testimony as fraudulent. When science knows enough about LSD and related substances—which it certainly does not at the present moment—it is likely that we will all be the beneficiaries.

There is hope, too, that fairly simple substances, applied judiciously, might help us learn and remember better and faster. "There is evidence," says Dr. Elkes, "that the injection of small amounts of substances which are of an excitatory nature, like strychnine, or small doses of picro toxin, or a drug known as physostigmine (which leads to the accumulation of acetylcholine) improves the rate of learning in animals, particularly in tasks of visual discrimination for food." More than that, animals who are exposed to a learning situation and *later* injected with one of these substances also learn faster. This is known as latent learning.

"Yet another indication," says Elkes, "comes from the effect of simple metallic ions (applied in minute doses) on the learning process. There is, for example, evidence coming from Dr. Roy John's laboratory that simple ions like potassium and calcium may affect learning. Small amounts of potassium, injected directly into the brain substance and allowed to circulate in the fluids which bathe the brain, enhance the learning process; inversely, calcium injected in a similar situation diminishes the learning process very strikingly."

But the expectation of really radical improvements in human abilities to learn, remember, and create, lies in a whole series of revelations about the essential nature of the brain's electrochemical storage system—the next subject to be taken up. When the fruits of this research have been harvested, man will truly transcend himself.

A man carries his history around in his head. He calls it his memory. This kind of history is anything but perfect or totally reliable, to be sure. But unless he has been a diligent and accurate diarist, or unless he has been of sufficient celebrity to induce others to keep records of his doings, then memories— his own or those of his family or friends—are all he has to fall back on. What is not remembered is lost.

Without the faculty of memory, a man could not be human. Without it there could be no learning, no education, no culture. Nor could there be any consciousness of the past, hence no more than a negligible ability to savor the present or to contemplate the future.

There may be those who would consider such a prospect refreshing. A surprising number of people do seem to be afflicted with an overwhelming ennui. Many of these emotionally and sensually poverty-stricken individuals must retreat into a drugged state to goad their flagging sensibilities into some sense of aliveness. (They manage to rationalize this failing into a sense of superiority.) Such people might welcome a world where every stimulus that impinged on the senses was brand new all over again every time it occurred. But what an unimaginably chaotic and insecure existence it would make for

241

the individual organism—one couldn't call it a man, or even a very viable animal—an organism that would in any case lack the capacity to appreciate the permanent novelty of a world where everything might be dazzlingly marvelous but nothing was ever familiar. In such a world every experience, every relationship, would be as evanescent as an ocean wave that leaps to a quick crest and dashes itself into spumed oblivion an instant later.

The essence of a memory is that it can be evoked without the necessity of repeating whatever experience or stimuli were originally responsible for it. Take that ocean wave, for instance. Though it has no memories of its own, we who see that ocean wave break on the shore can see it again in our mind's eye whenever we choose, long after it is gone, and recall the thrill—or, if we got wet, the chill—it may have given us. Since we live intimately with the phenomenon of memory, we tend to take it for granted. True, we get annoyed on those occasions when something is "on the tip of the tongue" and refuses to come the rest of the way. We may even worry that our memory is slipping. We all wish, in any case, that we could remember more dependably. We may even succumb, now and then, to a book or a course that promises helpful tips on how to improve the memory. We are frustrated by the certainty that all sorts of facts and experiences that remain stubbornly inaccessible to our gropings are stored somewhere in the disorganized filing system of our mind. Somewhere, but where? And how might we get better access to the things we want on instantaneous demand?

The mechanism of memory, "in which a brief experience may be transduced into a long-time functional change within the brain" (the words are those of Dr. Horace W. Magoun of U.C.L.A.'s Brain Research Institute), has always been one of the most elusive mysteries of the mind. "The physiological condition for memory and hence for learning," wrote the late

Dr. Norbert Wiener of M.I.T., "seems to be a certain continuity of organization, which allows the alterations produced by outer sense impressions to be retained as more or less permanent changes of structure or function." The human body and brain certainly have the requisite continuity of organization, and biologists have long assumed that there must be some kind of "engram" or memory trace recorded in the human brain or nervous system to make recall possible. But until recent years the guesses about the nature of the engram had little more evidence to back them up than did the primitive hypothesis of Socrates, who said: "There exists in the mind of man a block of wax and, when we wish to remember anything, we hold the wax to it and receive its impressions as from the seal of a ring. We remember and know what is imprinted for as long as the image lasts, but when it is effaced, or cannot be taken, then we forget and do not know."

Prior to the 1920's, knowledge of the brain's workings could only be inferred from subjective testimony or the observation of gross behavior patterns. But with the advent of electroencephalography, and later ESB, it became possible to measure the brain's electrical activity under a variety of circumstances. Studies of the brain's chemistry proceeded simultaneously, but for a long time the results were too confusing to contribute much to the understanding of learning or memory. The favored theories of memory regarded it, until very recent years, as an entirely electrical phenomenon. The late Dr. Karl S. Lashley of Harvard, among others, thought of memories as reverberating electrical circuits originated by nerve impulses from the senses and laid down in more or less permanent patterns that involved a great many of the brain's nerve cells. Lashley was unable to localize specific memories in specific areas of the brain. In fact, his experiments showed that memories tend to be diffused over extensive regions of the cerebral cortex. The reverberating-circuit theory was described most lyrically by the late British biologist Sir Charles Sherrington, who likened the brain to "an enchanted loom where millions of flashing shuttles weave a dissolving pattern." Could the electrical theory of memory account for the estimated million billion bits of information that a man's brain is called upon to absorb during his lifetime? Yes, it could, if certain assumptions were made about the complex interconnections among the brain's ten billion nerve cells (neurons) and their ninety billion supporting cells known as "glial" cells—cells heretofore unmentioned. Experiments finally made it clear, however, that the electrical theory alone was insufficient to explain memory. Those studying the electricity of the brain and those studying its chemistry realized, by the 1950's, that they could no longer pursue their studies in isolation. Brain activity, including memory and learning, had to be electrochemical in nature.

It is now generally—though not universally—agreed that memories start with electrical impulses and patterns, and that short-range memories are largely electrical in nature. But lasting memories, at least, must register themselves somehow in the chemical structure of the brain's cells. If you teach a rat a new trick, then immediately give him a strong electrical shock, you effectively knock out any memory of his newly acquired skill. If you wait fifteen minutes or a half hour before you shock him, his memory of the trick will be faulty but not altogether forgotten. If you wait a full twenty-four hours, the shock will have no effect; the rat will be able to perform as if you had administered no shock at all. The assumption here is that, for short periods of time, the rat's knowledge, learning, memory—call it what you will—was in the form of an electrical current, easily destroyed by the shock, but that later the continuing electrical reverberations had somehow insinuated themselves into the cell's chemistry, where they were safely stored and no longer vulnerable to any disruption of the electrical circuitry.

Other kinds of tests with animals showed even more convincingly that memory could not be exclusively electrical in nature. These tests consisted mainly in knocking out all the animal's electrical activity—by potent drugs and electroconvulsive shocks, or by deep-freezing the animal—and ascertaining that, after recovery from these traumatic experiences and the restoration of normal electrical activity, the former memories and skills remained intact. The purely electrical theories, then, as Dr. Lorente de Nó points out, cannot explain "how memory can persist after severe shock or deep anesthesia, i.e., after the circulation of impulses through a large number of cortical chains of neurons has stopped." In hibernating animals, says Dr. Hudson Hoagland, "cortical activity, electrically recorded, ceases entirely and cannot be elicited by electrical stimulation, however strong. Nevertheless, hamsters awakened from such sleep are able to perform previously learned tasks just as well after hibernation as before. Enduring memory thus appears to be based on some enduring chemical changes in the submicroscopic compartments of the neurons."

One of the most fascinating occurrences in that seminal period of the 1950's was the discovery, purely accidental, that specific memories are indeed stored in the brain and can be tapped repeatedly by means of ESB. Previously this sort of experience had occurred, and with much less predictability and consistency, only under hypnosis.

Dr. Wilder Penfield at the Montreal Neurological Institute was performing brain surgery on a woman with epilepsy, using delicate electrical probes to seek out the sites that might be involved in the patient's affliction. He was astounded to discover that when he stimulated a certain site in her brain, she could suddenly recall all the details of giving birth to her baby, as though reliving the experience. In another instance, when Dr. Penfield stimulated a certain site, the patient could suddenly hear music. Every time he stimulated the same site, the patient could hear the same music. Stimulating another area caused the patient to recall a past experience. Extending these experiments to other patients, Penfield was able to tap a variety of memories of the past, as far back as childhood, events the patient had considered long forgotten but now called up vividly, in great detail, with great accuracy and consistency, and evocative of the mood and emotions experienced at the time. "No man," said Penfield, "can, by voluntary effort, call this amazing detail back to memory."

"And so it is with our own past," wrote Proust. "It is a labour in vain to attempt to recapture it: all the efforts of our intellect must prove futile. The past is hidden somewhere outside the realm, beyond the reach of the intellect, in some material object (in the sensation which that material object will give us) which we do not suspect. And as for that object, it depends on chance whether we come upon it or not before we ourselves must die." This comes just before the famous passage in *Swann's Way* where the taste of a certain kind of cake soaked in tea is the instrument for bringing back to memory an entire period of the narrator's life. Had a Dr. Penfield been on hand, and if he could have applied ESB at the proper sites, this act might have replaced the "material object"—and, since ESB is repeatable at will, the calling-up of the memories would no longer have been by chance.

This ability to evoke a specific memory at will from a specific site appeared, for a while, to refute Dr. Lashley's notion that there were no such memory sites. But soon Penfield's own researchers convinced him that memories were indeed diffused over large areas of "the interpretative cortex" in a way that had led Lashley to declare ruefully in one of his papers that "learning just is not possible."

The widespread nature of memory storage was further confirmed by split-brain experiments in animals, carried out by Dr.

R. W. Sperry at Cal Tech, which provided convincing evidence that either half of the brain, when cut off from the other, manages to get along pretty well on its own.* There was really no reasonable way to explain this crazily dispersed storage system until molecular biology came to the rescue. With the elucidation of the molecular basis of genetics, it became clear that the large and complex DNA molecules in the nucleus of a fertilized egg cell could hold enough information to spell out detailed instructions for the entire development of the organism, information that would take up several twenty-four-volume sets of the Encyclopaedia Brittanica. An almost equally impressive molecule was RNA. And RNA soon became a prime suspect as the transmitter, if not the actual container, of

---

* Dr. Dean E. Wooldridge, fascinated by the split-brain experiments, has built upon them to suggest speculative experiments that shed some light—and some new question—upon the identity problems due to surgery taken up in Part I. If you take a given brain and call it Tom and sever the two hemispheres, "we will then have two Toms instead of one: the two half-brains will have exactly the same memories and each will feel itself to be the original individual. Here will be the first case of twins really entitled to the adjective 'identical.' And clearly we will be able to say correctly that, in this instance a new personality has been created by surgery." If we were able to reconnect the two hemispheres again, "two previously separate personalities will have been blended into one."

He then speculates upon what might happen if you took Tom and Dick, split their brains, and were able to put one of Dick's half-brains together with one of Tom's. "To be sure . . . the combination might not work. The differences in the patterns of thought of the two individuals might result, upon interconnection, in electric currents and chemical reactions preventing any kind of coherent performance of the hybrid organ; the combination might even be fatal to both half-brains. But if the combination *did* work, our theory would predict an interesting new kind of blended individual, possessing in one single consciousness the memories, learned habits, and senses of identity of both Tom and Dick. This would be schizophrenia with a vengeance!"

247

memory. Suddenly a hypothesis put forth by T. H. Huxley in 1874 seemed startlingly prophetic. "Each sensory impression," he had written, "leaves behind a record in the structure of the brain—an 'ideagenous molecule,' so to speak, which is competent to reproduce, in a familiar condition, the state of consciousness which corresponded with the original sensory impression. It is these 'ideagenous molecules' which are the physical basis of memory."

Using RNA as the theoretical "ideagenous molecule" of memory, it was finally easy to see, without stretching at all, that in this form the brain's cells would have the capacity to hold those million billion bits of information and more. Was it possible that the initial electrical circuits eventually caused structural changes in the molecules inside the neurons? Ingenious experiments and computer studies* by Dr. W. Ross

---

* The use of computers in conjunction with simple brainwave (EEG) tracings—i.e., with electrodes not implanted but merely in contact with the skull surface—has yielded some intriguing insights. For instance, a person will respond to a certain experience with a characteristic set of brainwaves. The EEG squiggles that appear when, say, a subject in the laboratory sees a light flash or hears a click or registers uncertainty will be roughly the same for one person as for another. Yet each person's squiggle will be unique enough to be thought of as his own individual "signature." Dr. Samuel Sutton, of Columbia University and the New York State Department of Health, recently expressed the awe he felt "when I would hand over one of these squiggly lines to an assistant and find that she was able to say: 'Oh, that's Jo Roberts when she makes a mistake,' or 'that's Dorothy Jacobs when she sees a flash of light and she knew ahead of time that it would be a flash of light." Similarly, Dr. Manfred Clynes of the Rockland State Hospital at Orangeburg, N.Y. can read in EEG tracings many reactions—for example, that what is happening at a given moment is that a man is looking at a red square, or a green circle—and believes he has gained deep insight into the mechanisms of perception. Dr. Rafael Elul of U.C.L.A. hopes that it may become possible to tell with certainty whether a person is telling the truth or not by applying computer analysis to EEG tracings—much as space scientists

Adey and his associates at U.C.L.A. indicated that this is what in fact happened. And it made sense in terms of RNA, which could occur in a sufficient variety of forms and combinations— easily a million billion of them—to make it perfectly capable of handling any memory load the brain might need to carry. And RNA was certainly widely enough dispersed. There are 20 million or more RNA molecules in every one of the brain's neurons, and these RNA molecules can manufacture some 100,000 varieties of protein.

In cell metabolism the chain of command, oversimplified, goes like this: DNA is boss. It issues the genetic instructions and RNA carries them out, largely in the form of protein building. So RNA would predictably be found busiest and most plentiful in cells that were growing, multiplying, secreting hormones, and the like. The surprise was that RNA turned out to be even more plentiful in the brain's neurons, which neither grow, nor multiply, nor produce hormones or any other secretions. Such quantities of RNA would hardly be just sitting there doing nothing. It might very well be the stuff memory was made of. To investigate this suddenly plausible possibility, a number of experiments were carried out.

---

analyze radar data to select from the welter of data only the significant blips that show, for instance, that a radar beam has been bounced off the moon. EEG studies have shown up significant brainwave differences between normal subjects and schizophrenic patients and have been used to study some of the mechanisms of hypnosis. "If the language of the brain is contained primarily in the activities of single cells and their interconnections," says Dr. Sutton, in *Bulletin of the Atomic Scientists,* "then the technique [EEG analysis] will probably run dry. It is simply revealing fortuitous correlations. . . . On the other hand, if the language of the brain is based on the joint activities of large aggregates of cells, with the activity of any single one being of minor importance—like the effect of the rise and fall of any single stock used in the Dow-Jones averages—then we can expect to forge ahead very rapidly with the available techniques."

The pioneer RNA-memory research was done by Dr. Holger Hydén at the University of Göteborg in Sweden. It took him a long, painstaking year to work out the marvelously refined skills and techniques he required to achieve his purpose: the extraction and isolation of single, individual brain cells so he could study them in their living state and analyze the changes wrought in the brain's chemistry by the learning process. He put his animals through a gamut of exercises, simple and complicated. For example, he made rats learn to do a tightwire-walking act in order to get to their food. Analysis of the neurons immediately after such training showed that the cells' RNA content had increased considerably—up to 12 percent more than in rats who had been permitted to quietly mind their own business. Not only was RNA present in greater quantity; in small amounts of it, the quality had changed as well. The chemical structure and composition of these molecules were different from any that appeared in the untrained control animals—a difference Hydén attributed to the training. Hydén's sophisticated measurements were even able to detect differences in RNA structure between the early and later phases of learning.

All this seemed to point fairly definitely to RNA as the carrier of memory via the chemical coding of its molecular arrangement. Hydén knew this could not be the whole explanation. DNA was probably involved, somehow, as the master activator of RNA; and the RNA perhaps used its encoded data as a means to build the new memories into protein molecules. But RNA's key role, as corroborated by other research, seemed undeniable. Meanwhile Hydén's research had disclosed the new importance of the neglected 90 percent of the brain's cells—the 90 billion glial cells (they are very sticky and their name comes from the Greek word meaning glue) which heretofore had been thought of merely as supportive tissue for the

10 billion neurons. What he observed was that, as RNA increased in the neurons, it diminished in the glial cells. It looked as if the glial cells supplied the neurons with RNA and other compounds when it was needed for learning purposes, then renewed their own supplies later. The glial cells now appear to have other, vital roles in brain function, and have become the objects of intensive and fascinated study.

Let us now consider what these experiments might mean. They could mean that a merely increased quantity of RNA added to the brain's cells might enhance the learning process or help improve memory. If this were so, then any substance that speeded up the cell's own rate of RNA production might have similar effects. Conversely any substance that inhibited RNA production would slow down learning and induce forgetfulness. On the other hand, if it was the protein that was crucial, and if there were a substance that could impair RNA's ability to make protein without otherwise affecting the RNA itself, a shot of *that* ought to inhibit learning and dull the memory. Over the past several years, all these experiments have been tried—and with interesting, if still controversial and inconclusive, results.

Yet more controversial are experiments carried out to explore the further implications of the research in memory molecules. The essence of memories is that they can be called up with a high degree of consistency. Therefore memory molecules, if they existed, would have to be consistent in their structure. And if they were, one way to prove it would be to transfer them—and, with them, their educated content—from one animal to another. In brief, if Einstein's understanding of relativity resided in certain RNA molecules of his brain, then the mere transfer of these molecules to another brain—if this could be achieved cleanly and without damage either to the molecules or to the brain—should bestow upon the recipient the

251

master's own thorough grasp of the subject. A partial transfer should transfer a partial understanding, and should make it easier for the recipient to learn the rest.

I will briefly describe this second set of experiments first. The best known of them—partly because the experimenter, Dr. James V. McConnell of the University of Michigan, is such a wittily articulate character—involve the cross-eyed flatworms called planarians, which McConnell has helped make popular in his highly readable magazine, *The Worm Runner's Journal.* The planarian can be taught by means of electroshocks to cringe whenever a light is turned on, or to avoid parts of a maze. When well-trained planarians were chopped up and fed to untrained planarians, McConnell observed that the cannibalistic worms were able to learn the same skills much faster than those on a less educated diet. (McConnell was thenceforward to be known in some quarters as "James V. Mc-Cannibal.")

In an attempt to find out whether or not RNA might be the transmitting substance, another experiment was done. A planarian, when cut in half, regenerates itself into two new flatworms. The researchers had already confirmed to their own satisfaction that when a trained planarian is cut in half, each of the new flatworms gets some of the benefits of the training. Knowing this, they cut a trained planarian in half. The top half developed, as expected, into a new planarian, partly trained. But the lower half, treated with a substance that destroys RNA, became a worm with no training—hence no memory—at all.

One of McConnell's associates in these experiments was a psychologist, Dr. Allan L. Jacobson, who moved on to U.C.L.A. where he soon graduated from planarians to rats and hamsters. He trained rats, injected material from their brains into untrained rats, who then speedily learned the same tasks. The only tasks they did learn with such unusual rapidity were precisely those that appeared to have been transmitted. Next, Jacobson and his colleagues reported success in transferring learning in a similar fashion from rats to hamsters and vice versa. At the Baylor University College of Medicine, Dr. Georges Ungar experimented with equal success in learning transfer. But these results have been challenged by other scientists—a prestigious list of them in universities, pharmaceutical laboratories, and the National Institutes of Health—who have tried without success to repeat Jacobson's (and, in some cases, McConnell's) experiments.

Nevertheless, there does seem to be tantalizing corroboration from other sources that the RNA molecule is somehow involved, if not central, in the structure of memory. Dr. Hydén's rats, for example, learned much faster during periods of high RNA activity. Dr. D. Ewen Cameron, now with the Veterans Administration Hospital in Albany, New York, started experimenting with RNA in elderly psychiatric patients while he was still at McGill University in Montreal. He first reported in 1958 that the memory of patients whom senility had not yet altogether overtaken seemed to improve when they were given RNA extracted from yeast.

These initial findings of Cameron's are hard to evaluate because, for one thing, they depend a great deal on subjective testimony—and, for another, it seems reasonably certain that RNA taken into the digestive tract is broken down there, and no one can yet be certain in what form it reaches the cells—especially the cells of the brain, which are guarded by the blood/brain barrier. Human beings cannot readily eat their education, as planarians, with their simpler biological systems, seem able to. There needs to be a method for ascertaining that RNA gets to the brain cells as RNA. The next best thing would seem to be a device for stimulating the cell to increase its own supply of RNA.

This is the approach being tested by Abbott Laboratories

with a pill called Cylert (magnesium pemoline). What it does—as revealed by tagging it radioactively and tracing its behavior—is stimulate a brain enzyme vital to the manufacture of RNA. Enzymes do not like to lie around idle. The assumption is that if you produce more of these enzymes, they will get busy producing more RNA. Does more RNA equal better memory and learning capacity? In rat experiments, Cylert rated an A-plus. The rats on Cylert not only learned much faster than normal, but retained their lessons for much longer periods of time. Now Dr. Cameron and others have been getting encouraging results with human beings—though it is still too soon to hail Cylert as "the memory pill," as some have been doing (though not its cautious manufacturer).

RNA's role in memory was confirmed by Dr. Gerard, who is now at the University of California at Irvine, when he reported in 1962 (while still at the University of Michigan) that by chemically speeding up the production of RNA in rats, he was able to speed up their learning; and, by chemically slowing down RNA production, to slow down learning. To demonstrate, though, that RNA was not itself the whole story, scientists at the University of Pennsylvania and the University of Michigan used puromycin—an antibiotic which, though it does not impair RNA in any other way, does interfere with RNA's ability to manufacture protein. The fact that puromycin also slowed learning and erased memory indicated that protein was indispensable to memory. Dr. Ungar's research at Baylor led him, too, to the conclusion that protein was the key. "These findings, however," he hastened to add, "are compatible with the assumption that RNA assures the continued production of the coded protein molecules."

Many brain researchers consider it a mistake to think of either RNA or protein molecules as memory *storage* devices. To them, the storage concept is too static, implying that the molecules themselves *are* the actual memories, or at least contain the explicit information recorded in their substance. Dr. Samuel Bogoch of the Boston Research Center, for instance, believes that the added sugar proteins present in the brain cells of his trained pigeons (versus untrained pigeons) are not in themselves stored information, but rather serve as switching mechanisms which somehow influence the electrical behavior of the neurons. Dr. Heinz von Foerster, too, emphasizes the dynamic nature of the process, comparing the molecules to tiny computers that respond to electrical stimuli. Thus, when molecules change their configuration as a result of experience, the change represents an *altered state of readiness* to respond in an altered manner to a given electrical stimulus, and to influence the neuron to behave in an altered manner. It takes *the whole process,* not just the molecular change, to make up the altered learning pattern. In some cases the neuron may act alone, in other cases with a few neighboring neurons, and in yet others in concert with a vast, interlocking array of neurons.

This more complicated picture renders more difficult the feat of transferring memory by transferring molecules. But the added difficulty certainly does not rule out the feat. It may be true, as Dr. Karl H. Pribram, Stanford's distinguished brain researcher, points out, that seeking the content of memory in the physiology of the brain is like "looking for the difference between jazz and symphonic music by studying the bumps on a record." But the bumps on a record do indeed constitute the material structure which, in contact with the needle, produces the music. The bumps are not the music itself, but they are, in a sense, the physical manifestation of it, since their configuration is what dictates that the electrical impulses will play back only the music "stored" in the record's sound track.

Hardly anyone denies that physical changes do take place in the neuronal molecules; the debate concerns only what the changes signify. Whatever the detailed truth turns out to be, the molecular basis of memory seems on its way to secure acceptance. So does the fact that training can be speeded up or slowed down, and memory improved or impaired, by chemical manipulation.

Improved intellectual capacity, then, is one of the major promises held out by the new age of biology. Dr. Rostand has suggested that we might increase the actual physical capacity of the brain by manipulating the embryo *in vitro*. Figure out a way to make the cells continue to divide just one more time—and, presto, the brain has doubled in size. A more likely possibility—and one closer to us in time, and available (as Rostand's solution would not be) to those already beyond embryonic manipulation—is one suggested by Dr. Francis Otto Schmitt of M.I.T., one of the world's leading brain investigators. A built-in misfortune of man is that, after the age of thirty-five or so, the cells of his brain begin dying off at the rate of 100,000 a day—and brain cells are not equipped, as other body cells are, to replace themselves. This adds up to a loss of some 36.5 million neurons every year. Even with a total of 10 billion to draw from, this is not a negligible quantity, and must certainly be an important reason why mental acuity diminishes and memory blurs with advancing age. Even in those neurons that remain, Dr. Hydén's experiments showed that the RNA content levels off in the forties and fifties (after steadily increasing from infancy to about age forty). As a man begins pushing sixty, he is not only losing those 100,000 neurons a day, but the remaining neurons begin to lose some of their RNA. Replacing the diminishing stocks of RNA, or stimulating its production, seems a distinct possibility. But more than that, Dr. Schmitt predicts that "chemotherapeutic measures, if discovered, may prevent the wholesale death of neurons and hence prolong the competent life of man."

As scientists and physicians expand their explorations in psychopharmacology, there seems little question that we all stand to benefit from having the electrochemical machinery of our minds geared up for quicker and easier learning, and better retention of what we have learned. This will probably be especially true in the case of infants and children, whose malleable minds are in any case laying down lifelong patterns.

Even more radical changes than those suggested by Dr. Schmitt can easily be envisioned. If a man learns French, algebra, or any piece of standard subject matter, what does this learning mean in terms of the brain? It appears to mean that the RNA molecules in this man's brain cells have recorded it (perhaps by transferring the coded knowledge to protein molecules) within their very structures. Remember Dr. McConnell's planarians, who seemed to be absorbing specific knowledge in the form of RNA from the educated planarians they supped on? And Dr. Jacobson's transfer of learning from rats to hamsters by means of RNA transfer?

Let us forget for a moment that these experiments are controversial and assume the transfer of learning via RNA molecules is possible. Suppose we learn precisely what the chemical structure of RNA molecules must be in order to structure them so that their codes will contain the knowledge we call the French language. Can we perhaps not create these molecules synthetically one day? Even if we possessed such RNA, there would of course still be the problem of getting it to the brain, and inside the cells. One of the reasons Dr. Cameron's experiments with yeast RNA are unsatisfactory is that the RNA had to go through the digestive system, where it was probably broken down. Could it be injected directly into the bloodstream instead? And could it be sneaked into the cells intact? Such speculations have led to the concept of artificial, beneficial viruses. (This is also one of the possible future methods—discussed in Part I—of combating the aging process.)

A virus is nothing more than a core of nucleic acid molecules, often RNA, surrounded by a coat of protein. The RNA

—if it is an RNA virus—is what contains the coded data that permits a virus to "infect" a cell. The way a virus works is to get its information, via its RNA, into the cell, to the cell's detriment. Now, if we had RNA that contained French, or algebra, and could make a synthetic virus out of it, it could be injected into the bloodstream. If we knew further how to get it past the blood/brain barrier (and certainly this knowledge should be in hand long before we can make synthetic RNA, specifically coded to order) and "infect" the cells with it, then we would be able to learn French, or algebra, or anything else whose code we knew, by injection. One can imagine education by mass inoculation, or the use of bacteriological warfare techniques for beneficent purposes by spraying entire populations with "good" viruses.* The teaching of many subjects would have thus become obsolete.

In France, Dr. Alexandre Monnier and Dr. Paul Laget have carried their speculations a step farther. If we know these nucleic acid codes, why not just write them into the original genetic specifications? Ridiculous? Look at the ants—they have an entire social system built into their genes. And look at the birds—they know how to build nests without ever being taught. Suppose we simply incorporated a lot of basic knowledge—say, the ability to walk, talk, swim, do arithmetic, play a piano, read ancient Greek, or whatever—right into the nucleus of the original fertilized egg so that the newborn would possess all these abilities without having to learn them?

Scientists never seem to offer a hope, though, without an accompanying warning. In this instance, Monnier and Laget suggest the possibility that "inherited knowledge, or even too swift and too perfect a memory, would prevent all further progress of our species.

---

* The same technique would of course be available to an enemy nation—or to one's own government—for the spreading of propaganda viruses.

258

"As far as insects are concerned," they point out, "evolution seems to have slowed down. These animals, which can do everything without apparently having learned anything, have hardly evolved at all for long geological periods."

Again the warning is clear: In seeking to control his own evolution, man could bring it to a standstill.

One thing that makes many of us feel discontented with the use we make of our brains is the fact that we see others— among them people who are not necessarily our intellectual superiors—making much more efficient use of *theirs*. We watch a TV quiz program, and witness an astounding performance by an individual—perfectly ordinary in every other respect—whose photographic memory enables him to store, and to retrieve apparently at will, encyclopedic quantities of information. We read of mentally retarded children, the so-called "idiot savants," who can do prodigious arithmetical calculations instantaneously in their heads. And we wonder, in frustration, what's wrong with us that we, with our higher I.Q.'s, can't do as well.

Sometimes a person who does not ordinarily possess such capacities can acquire them for brief periods while under hypnosis, or under the influence of certain drugs—and now by means of ESB. Subjects in deep hypnotic trances often have incredible powers of recall, even occasionally the ability to recite verbatim something once read aloud by a parent in childhood.* Apart from this unaccustomed access to the contents of their brains, subjects under hypnotic suggestion have been able to exercise unusual controls over their bodies, cre-

---

* Not that all one's lifetime experiences are necessarily kept in the memory. Dr. Heinz von Foerster of the University of Illinois, for one, rejects this notion. "If people stored all the nonsense they have ever seen," he believes, "they could never retrieve anything."

ating real burn lesions on the skin without being burned, causing warts to vanish, bypassing sensations of pain, and exhibiting other strengths and muscular capacities normally beyond them.

Calling up memories by ESB is in one sense quite similar to calling them up under hypnosis. In both cases the memories come forth through the intervention of an outside experimenter. But in ESB the experimenter must know *where* to stimulate before he can be sure of getting the memory he wants. In hypnosis the experimenter has no idea where anything is stored; all he does is make the suggestion, and *the brain does the rest all by itself*.

Whatever the biochemical secrets that govern these intriguing powers, whatever the chain of command from conscious through subconscious via nerves and glands and organs and muscles, these areas of investigation—once reserved for the occult—are now of great new interest to science. The long-range promise is that these powers, instead of being sporadic and conditional, could be permanent and constant, and that our own conscious minds could replace the outside experimenter. Think what our intellects might be like if we could really remember virtually everything, and have all the significant data stored away for instant retrieval. Think of being able to control many of our appetites, sensations, and physical abilities by giving conscious orders through the subconscious (whatever that turns out to be), orders to be carried out by those organs and systems of the body to which no one—except perhaps for certain practitioners of Yoga—now has any direct access at all.

No one really knows what the capacity of the human brain might be, used to its limit. But that limit, whatever it is, may soon be overcome by hooking up the human brain to a computer. When this kind of brain-computer hookup is imagined, it is usually to envision the computer as remotely controlling and directing the brain—or a dozen brains, or a hundred. There is no reason why it could not work the other way around, with the computer serving as a vast storehouse of readily accessible information for the brain's use. ("Intellectronics" is what Dr. Simon Ramo calls this new field of study.) "A man could get himself hooked up to a computer that has been programmed to teach painting in the style of Cézanne, for instance," says Dr. Kurt von Meier of U.C.L.A., "or to a machine that could teach him to write music like Mozart's. It should be possible to learn any skill or any art in this way."

We need not stop here. Some of the same experiments that revealed how specifically a man's memories can be tapped also revealed that a man may be made to remember things that never happened to him at all. A hypnotist, for example, can implant a totally fraudulent experience by suggestion, and the subject may awake and believe the incident really happened.

Now, we have just been speculating about the possibility of acquiring knowledge via injection, and how viruses of the future might give us French or algebra instead of the flu. All this was based on the assumption that the details of memory and learning are built into the chemically coded structures of the molecules inside the brain cells. These molecular changes, however, are originally brought about by electrical impulses. Memory, as we know it, seems to be acquired electrically, though stored chemically. So it seems feasible that memories could be implanted electrically if we knew the mechanism.* Arthur C. Clarke is convinced that the famous "mechanical educator" of science fiction—the machine that "could impress

---

* Controlling the brain by means of electricity from the outside is a fairly common fancy. But there is no reason why the brain's own electricity—especially if a practical amplifier can be devised and appropriate hookups made—might not be used to control things outside the body. A marvelous prospect for the lazy.

on the brain, in a matter of a few minutes, knowledge and skills which might otherwise take a lifetime to acquire"—is now no longer necessarily relegated forever to fiction. He believes information might be fed into the brain almost as sounds are recorded on magnetic tape, to be stored there for playback on command.

If true information could be recorded and stored this way, why not false information? And if information is recordable as on a magnetic tape, might it not be erasable as well? (The erasing—which might have to be done chemically—could be considerably more difficult to achieve.) Any knowledge no longer useful, any memory of an experience a man would rather forget, could be wiped out as if it had never been there at all. And artificial experiences could be supplied at will. A man could rewrite his own history—or have it rewritten for him (perhaps without his knowledge), a form of literal brainwashing—just as the totalitarian nations have often rewritten theirs. With such powers and practices loose in the world, who could ever be certain that his memories and experiences were real, and not implanted? If people, already rendered totally docile, could be programmed simply to plug themselves in at a certain hour of the day—or turn on their receivers, or whatever the technology of the day calls for—dictators who controlled a few communications satellites in the sky could transmit thoughts, moods, feelings, experiences, even personalities to entire populations. In the new age, Pontius Pilate's question would take on horrifying dimensions.

A search is already under way for methods to bypass the senses through electronics. We see and hear, not with our eyes and ears, but with the visual and auditory areas of the brain to which signals are transmitted by the eyes and ears. If devices could be invented—and some primitive prototypes already have been—that could scan the outside world and transmit the same signals to our brain, we could see and hear without eyes and ears. If ways can be found to bypass the senses, ways may also be found to expand them into areas of sensation and experience far beyond the threshold of our current biological sensors, and even to provide us with the senses which we can now no more imagine than a mole can imagine color.

Sights, smells, sounds, and other sensations have been artificially produced through ESB, drugs, and hypnosis without the presence of anything "real" to see, smell, or hear. "Artificial memories, if they could be composed, taped and then fed into the brain by electrical or other means," says Arthur Clarke in *Profiles of the Future,* "would be a form of vicarious experience, far more vivid (because affecting all the senses) than anything that could be produced by the massed resources of Hollywood. They would, indeed, be the ultimate form of entertainment—a fictitious experience more real than reality." If a man could experience all the sensations of making love to a beautiful woman, and even remember it vividly as though it had happened, might he not prefer this to the arduous, anxious and expensive pursuit of a real flesh-and-blood female?* Would he go to Egypt to see the Pyramids if he could feel, see, smell them in all their rich splendor right at home? "It has been questioned," says Clarke, "whether most people would want to live waking lives at all, if dream factories could fulfill every desire at the cost of a few cents for electricity." If the citizens of a technically advanced nation ever succumbed to such temptations, they might easily fall prey to a relatively barbarian nation whose people still liked its experiences raw— or a nation that employed its technical resources more aggressively and less sybaritically.

---

* Arthur Clarke himself once wrote a short story called "Patent Pending"; the patent applied for being on a gadget designed to tape sexual experiences in their entirety and playing them back whenever desired.

Clarke carries his line of reasoning yet further. Since the brain is the key to experience, then as long as the brain were left it would not matter how much of the body was replaced by artificial sensors—even if they finally were totally substituted for the body, to serve a stored and protected but actively sensing brain. Responsible scientists support Clarke in these imaginings. At a 1965 meeting, for instance, Dr. James Bonner, a distinguished Cal Tech biologist, was discussing the possibility of prolonging the life of the brain and of increasing its capacity by increasing its size. If brains thus got too big to carry around, Bonner suggested, what about "ones that will stay comfortably in one place and send their sense organs out into the world? We may develop new sense organs for receiving and sending microwave signals. That would be very convenient for communication at a distance." Such a living and sensing though bodyless brain was conjured up in Curt Siodmak's novel, *Donovan's Brain*. Would anyone not shudder at the thought of continuing existence in this manner? Dr. J. D. Bernal once wrote that, while such a brain would be "purely mental and with very different delights from the body," it might still be "preferred to complete extinction." Arthur Clarke thinks that it definitely would be. In fact, it might be preferred to the way we now live and experience. "If you think," says Clarke, "that an immobile brain would lead a very dull sort of life, you have not fully understood what has already been said about the senses."

Clarke's contention is borne out by a fascinating study, made by Dr. John Money of Johns Hopkins, of the sex lives of paraplegics. In a male paraplegic accustomed to having normal sex relations prior to his incapacitation, the former mechanisms of sexual behavior are no longer possible because (depending on the nature of his incapacity) the brain center is out of communication with the "erectile" center in his spinal cord. (It is this center that is normally "turned on" by the brain, and this turn-on triggers the physiological process of erection). In such cases, two things can happen separately: The patient, through local stimulation—say, manual manipulation of the penis—can trigger his spinal center so that an erection occurs— but, though he may look down and see it, he has no feeling or conscious awareness of it; so, in subjective terms, it is meaning- less, since it is *not really an experience.* Or he can, through fantasy or dreams, which take place in the brain alone, go through the gamut of sexual sensations, from erection through ejaculation—when in fact, nothing whatever is happening in his lower anatomy. The sexual action here is obviously all taking place in the brain.

Dr. Wooldridge has imagined experiments in which a hu- man brain might be kept alive on its own, much as Dr. White kept alive the monkey brain in Cleveland. It would be nour- ished and kept electrochemically "conscious," and Wooldridge further conjectures that it could be supplied—through elec- tronic devices—with perceptual capacity (such as TV-like sig- nals to the optic nerve) as well as some sort of speech mecha- nism. "Perhaps a word of reassurance is in order," he allows, "regarding the seemingly gruesome aspects of the postulated experimental arrangement. We do not necessarily have to feel sorry for the disembodied star of our planned production. In the first place, we will certainly control the conditions so that it feels no physical pain or discomfort. Furthermore, from the work on pleasure and punishment centers, we know that we can also control its emotional state, making it feel continually relaxed, happy, or even ecstatic simply by arranging for suit- able patterns of electric current in selected regions of the brainstem. Indeed, if such experiments ever really become pos- sible," he predicts, "a major problem may be the selection of lucky winners from the many who volunteer for disembodi- ment because of their wish to achieve a happier state of existence than that available to them by ordinary means."

A brain connected by wire or radio links to suitable organs could participate in any conceivable experience, real or imaginary. When you touch something, are you *really* aware that your brain is not at your fingertips, but three feet away? And would you notice the difference, if that three feet were three thousand miles? Radio waves make such a journey more swiftly than the nerve impulses can travel along your arm.

"One can imagine a time," says Clarke, "when men who still inhabit organic bodies are regarded with pity by those who have passed on to an infinitely richer mode of existence, capable of throwing their consciousness or sphere of attention instantaneously to any point on land, sea or sky where there is a suitable sensing organ."

To help himself evolve and explain his theories of relativity, Dr. Einstein made extensive use of what he called "thought experiments." These are experiments which might or might not be possible in the real world but which he could easily carry out in his imagination. To clarify his concepts about the nature of gravitation, for instance, he found it useful to imagine such scenes as a man traveling through space in an elevator without being able to tell whether he was going up or down.

The technique of the thought experiment has lately been used by Dr. Wooldridge to clarify the discoveries made in recent years about the electrochemistry of the brain, the molecular biology of genetics, and the relationship of computers and people. Two of these experiments—splitting brains and putting them back together; and equipping an isolated human brain with the power of speech—have already been described. His logical follow-through of where such experiments might lead casts considerable illumination on some of the dilemmas centering on personal identity, life versus death, the soul. In his book *Mechanical Man: The Physical Basis of Intelligent Life* (from which the two previous examples were taken), he imagines

> an experiment on a normal adult human being comprised of the following steps: (1) Employing suitable anesthesia, we connect appropriate apparatus to the subject's brain and make a permanent record of the complete contents of his memory (including physical and mental habits, of course). (2) We transfer this record to the neuronal material of a fresh, factory-made human brain. (3) We replace the natural brain of the subject by the newly processed factory-made article, making suitable connections to all incoming and outgoing nerves. (4) We allow the subject to awaken. The pertinent question is "What, or rather who, is the result of this series of steps?"
>
> To all outward appearances the individual is unchanged by the operation. His memories, habits, patterns of thought, skills, worries, fears, and satisfactions are exactly as they were before. These are all determined by the structure and chemistry of the material of the brain, all essential aspects of which have been accurately established in the synthetic organ by the postulated duplicating process. To be sure, the material of the brain is all new, and our conventional modes of thought almost automatically cause us to conclude from this that the old individual is gone. But is he, really? After all, metabolic processes continually replace the molecules in the normal brain; hardly any material has been with any of us for more than a few weeks. Yet we do not believe that this causes us to become new individuals. Logic, as well as the thesis of the machinelike nature of man,* would seem to require us to take a similar point of view with respect to the experiment we are considering—to conclude that the personality of the subject not only *appears* to be unchanged but *is* unchanged, despite the complete replacement of the material of his brain.

He then goes on to describe a related thought experiment dealing with interstellar travel.

---

* The machinelike nature of man is the major theme—and an arguable one, of course—of Wooldridge's book.

The problem posed by the relatively short life span of the astronaut is an obvious one. The solution, according to our line of speculation, could be as follows: Record the contents of the astronaut's brain and his DNA genetic specifications on durable, long-lived storage material. Arrange a mechanism in the spaceship which, at the future time of interest, will automatically combine suitable inert chemicals and subject them, under control of the recorded genetic specifications, to the physical manipulations involved in the man factory described earlier. Then, by an equally automatic process, impress on the virgin material of the new creature's brain the total recorded contents of the brain of the ancient astronaut. Finally, allow the finished product to awaken in time to meet the inhabitants of the planet to be visited. Pure science fiction though it may be, this imaginary sequence of events illustrates again one of the most bizarre consequences of the idea of the machinelike nature of man. For even though thousands of years should intervene between the recording of the design details of the astronaut and his subsequent duplication, we must conclude that it would be as accurate to call the two organisms the same individual as it is to call any one of us today the same person he was yesterday.

By similar thought experiments, Wooldridge makes it clear that these eventualities pose no real dilemmas for him in terms of life or death, identity, or the human soul. Believing, as he does, that "consciousness is only a transient property of the current state of organization and electrochemical activity of matter," he simply defines this consciousness as the individual's identity. (It follows that the consciousness of a *new* identity transferred to the biological material of an already-existing person would in fact make that individual a new person. The consciousness is what counts.) When this consciousness is operating, the individual is alive; when it is inoperative, he is dead. In the case of the hypothetical astronaut, it could be said that he was dead for those thousands of years during his flight through space, and that he was brought back to life again when the consciousness was reactivated. As long as the information for his reanimation was safely stored, perhaps it would be more accurate to say that he was in a state of suspended potential animation during the long interim. Only if the information were permanently destroyed could he really be pronounced dead.

And where was the soul meanwhile? To Wooldridge this is like asking, "Where does a light go when it goes out?" Just as the light is a phenomenon that exists when electricity goes through a certain kind of filament, and no longer exists when the electricity stops; so does the soul—or life, or identity—fail to exist when consciousness stops. Thus life, soul, identity, all are one and the same as the transient consciousness, and all these dilemmas are solved—if you accept Wooldridge's basic thesis. It is not easy to refute with any finality.

While the human brain is busy making itself ever more wonderful (though few of us can adopt Arthur Clarke's rapturous view of these eventualities), what about the competition?

Competition?

Yes. From the computers themselves, which we tend to refer to loosely as mechanical brains—a tendency deplored by many. If these man-made brains became advanced enough, why should they continue to work for us? Why not go into business for themselves?

Computer scientists and engineers, especially those who work for industrial computer concerns, never tire of pointing out the limitations of their product. Yes, yes, computers do have incredible talents. But, after all, they can only do what man programs into them. They are not, and never can be, capable of independent thought, we are constantly assured. Words like "memory" and "learning," applied to computing machines, are inapt analogies, and all such anthropomorphic terminology ought to be eliminated. What computers do is mechanical, not creative. Et cetera.

But all this negative prophecy about the future capabilities of computers is premature. The age of computers is barely in its infancy, and already their capacities have far outrun earlier predictions. Why the hurry, then, to set limits where these limits cannot be known? Many computer scientists, especially those in the academic world who neither make nor sell computers, but only use and design and study them, believe that judgment on these matters will have to be withheld. Dr. Wooldridge is among those who do not see why advanced computers might not be capable of real learning through experience, and of truly independent thinking. They will therefore be legitimately designatable as brains. Such computers may be made of ever more sophisticated and ever more miniaturized materials. They may even be made of organic matter. Dr. Kenneth Boulding has suggested that brain cells kept alive *in vitro* might be fine stuff for computers of the future. Even more efficient might be organic material totally synthesized in the laboratory.* Whatever they are made of, computers may very well possess what Dr. Bernice Wenzel of U.C.L.A. calls "extraorganismic intelligence."

"All computers can do *now* is put out what's been put in," Dr. Wenzel readily admits. "But at the present rate of progress in research in computers and intelligence, as well as in physiology and the other applicable sciences, I can't believe it will not be possible to create this 'thing'—this lump of artificial intelligence, this computer that won't have to be programmed at all . . . Eventually people may not be able to tell whether the 'thing' they're communicating with is a synthetic thinking machine or a human being." On that day, who can make the distinction between an artificial brain and a "real" one? Do such independent intelligences then become legal entities? Are they to be recognized as persons? Do they have rights?

Whether we think so or not, suppose *they* think so? And if

---

* "There must be few scientists left," writes Dr. Pierre Auger in the *Bulletin of the Atomic Scientists,* "who would assert that it is impossible for man to achieve the synthesis of living organisms."

we choose not to give them their rights, they might simply seize them. "There is no doubt," says Richard R. Landers in *Man's Place in the Dybosphere*, "that machines as a group will dominate man as a group, and eventually, individual machines will dominate individual men.

"By dominate I mean control, regulate, restrain, influence, pervade, direct, guide, prescribe, etc., by virtue of superiority in all aspects of tasks demanding a leader."

> Once we have devised programs with a genuine capacity for self-improvement [writes Dr. Marvin L. Minsky of M.I.T. in *Scientific American*], a rapid evolutionary process will begin. As the machine improves both itself and its model of itself, we shall begin to see all the phenomena associated with the terms "consciousness," "intuition," and "intelligence" itself. It is hard to say how close we are to this threshold, but once it is crossed the world will not be the same.
>
> It is reasonable, I suppose, to be unconvinced by our examples and to be skeptical about whether machines will ever be intelligent. It is unreasonable, however, to think machines could become *nearly* as intelligent as we are and then stop, or to suppose that we will always be able to compete with them in wit and wisdom. Whether or not we could retain some sort of control of the machines, assuming that we would want to, the nature of our activities and aspirations would be changed utterly by the presence on earth of intellectually superior beings.*

---

* In Stanley Kubrick's movie, *2001: A Space Odyssey,* the computer aboard a spaceship en route to Jupiter outwits and does in all the astronauts except one—who finally saves himself by sneaking in to dismantle the computer's brain centers.

In a slender science-fiction novel called *The Tale of the Big Computer: A Vision,* a well-known Swedish physicist writing under the pen-name Olof Johannesson has his computer-narrator of the future say in praise of man—by this time virtually displaced—that he is "the only living creature intelligent enough to perceive that the purpose of evolution was the computer."

In dealing with computers, as in dealing with so many of the potent new tools and techniques, man is very much in the position of those Japanese gastronomes who are addicted to the fugu fish. The fugu secretes a deadly poison to which there is no known antidote. Yet its flesh is considered so delectable that Japanese aficionados are willing to pay high prices for it and risk the dangers. The risk is somewhat mitigated by the fact that in Japan only licensed chefs, specially trained to know the poisonous parts from the nonpoisonous and to prepare it non-toxically, are permitted to serve it up. No system is foolproof, of course. Despite all precautions, a couple of hundred Japanese die every year of fugu poisoning.

Perhaps we should all take a lesson from the fugu fanciers. To enjoy the pleasures, we must take the risks. But let us by all means see to it that the chefs know what they are doing.

All experience, in fact everything that exists in the universe, is what it is because of the way its information is organized. This is one of the most profound and fundamental insights of contemporary science. An atom of carbon is an atom of carbon because of the explicit arrangement of its constituents. Re-scramble the arrangement and you may get many other things but never a carbon atom. The entire computer revolution and the resultant strides in automation flow directly from the development of information theory. Molecular biology, as earlier chapters emphasized, now tells us that our heredity and all the life processes are dictated by biochemical information stored in the cell nucleus. Change the information, change the organization of the atoms in the DNA molecule, and you get a different set of characteristics, even a different creature. Thus, as manipulation of the biochemical information that governs our bodies and brains becomes more sophisticated, so will our control over our bodies and brains. Carried to its theoretical extreme, this would mean that anybody who possessed the necessary information and the means to manipulate it *could do practically anything to practically anybody*.

Information can now be transmitted virtually instantaneously over long distances. As a matter of fact, information is just about the *only* thing that can be transmitted instantaneously. When we send a voice over the radio, it is not the voice itself we are sending but rather a one-dimensional signal which, when it goes through the receiving apparatus, is reconstituted as the sound of a voice. When we transmit television pictures, again we do not send actual pictures. We extract and record the appropriate information, this time in two dimensions—which is why television was so relatively slow in being developed commercially—then send the information and finally put it back together as an image on our receiving screen. (More information is needed for color TV pictures than for black-and-white.)

Now, suppose information could be transmitted in *three* dimensions—a possibility that fascinated the late Dr. Norbert Wiener. If we could somehow extract from a solid object *all* the relevant information, the incredible complexity of data on every one of its atoms and all their interlocking operations and relationships, then we could also record this information. And if we could record it, we could transmit it and reconstitute it at a receiving station.

It is theoretically possible, then, that three-dimensional objects, including people, could be teleported across long distances instantaneously—or at least at the speed of radio waves or laser beams. Thus, a man might be re-created not merely from his original chemicals—the long-dreamed-of creation of life from inorganic materials in the laboratory—but from the *raw information*.* And if the information were stored, then

---

* This is not unrelated to Dr. Wooldridge's thought experiment with the interstellar astronaut.

any number of copies could be made. This would be the most efficient people-raising method of all. Instead of keeping cells alive in tissue culture from which genetic duplicates could be grown over long years of care and nourishment, a man could just store himself on tape so that, if any accident befell him (including old age), a duplicate of himself *as a healthy adult* could be produced by playback.

But these are the fuzziest fancies of all, and even men with the most free-wheeling imaginations, men like Dr. Wiener, Dr. Wooldridge, and Arthur C. Clarke, admit that such eventualities lie, if anywhere, in the remotest future. Meanwhile, man-machine hookups are becoming very much a reality, and cybernetics—the science which Dr. Wiener founded—is thriving. To the brilliant new breed of cyberneticists, the technical problems of man-machine combinations are a routine part of their daily work. They can hardly be expected *not* to be hatching up futuristic schemes of their own.

Cybernetics* is basically the study of the relationship between computing machines and the human nervous system. It deals with the art of handling vast quantities of information, running the data through complex computing systems which then feed back new and useful data. To run an automated oil refinery, for example, information is constantly being absorbed and fed back in order to carry out all the intricate industrial processes involved. There are scientists who now believe that the human body could be run cybernetically—i.e., could be automated—in a similar fashion, though of course the job would be much harder. The proposed automated man is called

---

* Richard Landers of Thompson Ramo Wooldridge, Inc., has coined the term *dybology* (from the Hebrew word *dybbuk* meaning "unassigned soul") to define the "growing area between biology and engineering that cannot be classified as being strictly biological, or strictly engineering." This would cover the whole area of lifelike, life-supporting, or life-mimicking mechanisms.

a cyborg—for cybernetic organism. The cyborg, though cybernetically controlled, would be still a human being, though some might find it difficult to consider him that after such radical tampering.

A concept like the cyborg might ordinarily have languished in the limbo of sheer fantasy, but in the space age it happens that NASA has practical space-faring problems which even the partial realization of the cyborg could help solve. Hence this unlikely creature may be accelerated into some form of actuality much sooner than anyone might have foreseen.

The term cyborg has lately been applied loosely to almost any hookup between man and an artificial device—even a wooden leg or an iron lung. But as originally conceived by Dr. Nathan Kline and Dr. Manfred Clynes of the Rockland State Hospital in Orangeburg, N.Y., the cyborg implies a thoroughgoing modification of the human body. A cyborg designed for astronautics would still resemble a man, but an unearthly one indeed. He would be encased in a skintight suit, needing no pressurization because his lungs would be partially collapsed and the blood in them cooled down, while respiration—and most other bodily functions—would be carried on for him cybernetically by tiny artificial organs and sensors, some of them attached to the outside of his body, some of them implanted surgically. His mouth and nose, too, would be sealed over by the suit, because he would not need them to breathe with. Cyborgs would communicate with each other by having the electrical impulses from their vocal cords transmitted by radio. The artificial organs—a miniaturized computer system constantly receiving and feeding back information to regulate the body to its changing environment—would keep a cyborg's metabolism steady despite radical fluctuations in external temperatures and pressures. The cyborg could travel in an unsealed cabin through the vacuum of space, walk around on the moon or on Mars protected from heat, cold or radiation by a variety of chemicals and concentrated foods being pumped directly to the stomach or bloodstream. Wastes would be chemically processed to make new food. The tiny bits of totally worthless waste matter would be deposited automatically in a small canister carried on the back.

An earthbound cyborg would not need to be as elaborately fitted out as his space-going counterpart. He wouldn't need the space suit or the canister. Nor would he need the same kind of extraordinary protection from hostile surroundings. But the fact that NASA has put some of its money on the line for research toward the cyborg is an indication of the seriousness with which these possibilities are being pursued.*

Practicing one-upmanship on the cyberneticists, Dr. Michael Del Duca, a scientist formerly with NASA, predicts that miniaturization will proceed to the point where the control mechanisms can be molecule-sized—in effect, sophisticated chemicals which could pass through the skin in either direction, moving wherever needed to perform their multifarious wonders. Del Duca is confident, too, that man could ultimately learn to convert sunlight directly into energy, just as the plants do. Practicing photosynthesis, he would not need food at all. By then, too, the need for sleep will have been done away with—or at least curtailed to a small fraction of the one-third-of-man's-life now spent in this useless pastime. All this may be bad news for

---

* The practical concerns of space travel have brought on many a bizarre and unpredictable problem. As one example, NASA is quite seriously collaborating with the Public Health Service and the Department of Agriculture—and has even appointed a Planetary Quarantine Officer— to grapple with problems of how to decontaminate returning spacecraft and astronauts who have been to the moon or Mars. Just as American Indian tribes were decimated by diseases that were only minor annoyances to the Europeans who brought them, so could some alien microorganism, even if innocent and harmless in its own environment, devastate the earth.

gourmets and sluggards, for farmers and mattress manufacturers, but people like Del Duca and Clynes are certain the rewards will more than compensate for the deprivations. They envision the photosynthetic cyborg of the nth century freely moving anywhere—walking, levitating, flying, natating—through the earth's atmosphere, through the ocean depths, through the no-longer-hostile vacuum of outer space—anywhere in the accessible universe, free to spend virtually full time in exploration and in the pursuit of creative activities.

In those far distant days, if the prospects are to be believed, every man can be his own Shakespeare, Michelangelo, Beethoven, and Newton combined. In fact, what formerly passed for creative genius may seem puny compared to the shining technicolored raptures to come. The late Hindu poet and philosopher, Sri Aurobindo, praising the "life divine" which he believed would ensue from the new evolution, prophesied, "If mankind only caught a glimpse of what infinite enjoyments, what perfect forces, what luminous reaches of spontaneous knowledge, what wide calms of our being lie waiting for us in the tracts which our animal evolution has not yet conquered, they would leave all and never rest till they had gained these treasures."

Gaining such treasures will, in Robert Ettinger's view, be one of the principal motivations for the freezer program he so ardently advocates. Because he feels that research will soon improve freezing techniques quite dramatically, his first bit of advice to individuals who sense the approach of death is: Try to survive a little longer. We can do worse than offer to a harried and hard-pressed world the same good advice: Try to survive a little longer.

277

PART IV

Afterword

According to Pico della Mirandola, writing in the fifteenth century, we men are somewhere at the midway point of creation, capable of rising toward the angels or of descending toward the beasts. That was the end result of Genesis I. As the Second Genesis approaches, we find our powers to go either way enormously increased. The one thing we are not empowered to do is to remain at that midway point, frozen in place like the treasured, timeless figures on Keats' Grecian urn. We live in an evolutionary continuum, a dynamic multidimensional spacetime reality where rest is only a theoretical state. Either we choose to go forward—upward, in Pico's sense, toward a finer tomorrow; or we retreat toward bestiality—a plunge into a sinkhole where indeed no beast would venture.

Such a plunge toward bestiality has been the inevitable consequence of any period of drastic change, Eric Hoffer observes, because, in his view, such a period puts an intolerable strain on the cerebral cortex—the most lately evolved, and therefore the most human and most vulnerable, portion of the brain—leaving us at the mercy of the more primitive brain centers. Thus, he believes, any era of upheaval is bound to be a "time of barbarization." "Each generation," says Hoffer, "has to humanize itself."

To humanize ourselves—to retain the primacy of the cortex and its controlled intelligence; to refuse to turn our minds over to the emotion-wracked direction of our primordial brains; to bypass, somehow, the customary "time of barbarization" in the face of all strains and social upheavals—is the primary task of all the still-extant generations. To humanize ourselves is to brake our precipitous plummet into what could speedily become a retrograde evolution. Down, to be sure, is the easy route where inertia will carry us without effort (though not without painful consequences). Ultimately, then, to humanize ourselves is to start our minds and muscles moving—pushing against gravity, against entropy—in the direction of our next evolu-

281

tionary advance. With the help of BSP we can imagine a variety of post-civilizations that might grace Genesis II and a variety of trans-humans to inhabit them. We can make our choices—not precise choices, the limits of imagination and prognostication being what they are, but broad ones, enough to set our general direction. That done, our assignment is to figure out how we begin to get from Here to There. As our new powers develop, we should accept them, welcome them, and not let our caution turn to dread. We must trust ourselves to use the powers for our benefit and to find ways to avert the evils—perhaps employing our new controls for the latter purpose as well. We must, in a word, exercise "quality control" over our future.

As we eye the contemporary landscape, we see little to sustain an optimistic overview. We find it hard to fault Arthur Koestler for concluding, as he contemplates "the streak of insanity which runs through the history of our species," that "man's native equipment, though superior to that of any other living species, nevertheless contains some built-in error or deficiency which predisposes him toward self-destruction." He ultimately recommends, as the sole solution for incorrigible man, the development of a civilizing new Pill to subdue the baser instincts. But if we cannot find some early solutions by using our intellects to influence our attitudes and behavior, if the poverty of our vision leaves us with electrochemical auto-manipulation as the *only* hope, then we must indeed, in the words of Shaw's Dr. Barnabas, "go the way of the mastodon and the megatherium and all the other scrapped experiments."

Much of the anxiety in a world where man lives side by side in uneasy symbiosis with The Bomb stems from a prevailing lack of confidence in a viable future. It is sobering to realize that our civilization is after all a flimsy man-made thing which, if it now disappeared through man's own whim or folly, would have been but a brief transitory itch on the earth's skin, a bit of static quickly drowned out in the indifferent loud chaos of eternity.

In the face of this overwhelming realization, one major obstacle to a more sanguine outlook is our stubborn reiteration of defeatist clichés: there have always been wars, there always will be wars, you can't change human nature, etc., etc. If we believe that the human nature which can't be changed is a nature already fundamentally flawed, then our hopes are doubly crippled. Indeed, the gloom and despair in our society is due in large measure to the image of man we have permitted ourselves to buy: man, rotten to the core, doomed by his heritage to murderous aggressiveness—a creature therefore without hope, living a life of total absurdity in a nowhere society. I believe it is a false image—purchased on the skimpiest of evidence—and a yoke around our necks. Yet such is our self-image in the year A.D. 1969, one of the most interesting times in human history, with a fantastic era of change and challenge ahead.

The penchant for self-disparagement is certainly not unique to our day. Many prominent figures of history have held an equally unflattering estimate of man's nature and an equal pessimism about his future. If you believe—with the little man in cartoonist William Steig's little box—that people are no damned good, you will be supported by plenty of philosophers, going back at least to Thomas Hobbes. You will have the support, too, of Dr. Sigmund Freud, who wrote of the primary hostility of men toward one another. "If we are to be judged by our unconscious wishes," he said, "we ourselves are nothing but a band of murderers." He flatly declared that "the tendency of aggression is an innate, independent, instinctual disposition in man." Freud's judgment—for which, in my view, he never adduced convincing evidence—has more recently been upheld by the animal behavior studies of modern ethologists. But I think that too much has been too glibly extrapolated from animals to man on too slender a thread of scientific evidence.

Men are not ants or bees with their individual and collective futures rigidly preordained in their genetic lattices. Human genes set limits, too, but we are born with a spacious grab bag of possibilities, and we undergo a long stretch of leisurely development during which we remain malleable and subject to shaping by parents and teachers, by friends and enemies, by social patterns visible and invisible at all levels of consciousness. We are in a very real sense the products of our total cultural environment. We know from experience and from experiment that people and their surroundings are both alterable, and that altering one can alter the other. We *can* change human nature.

"Is the deliberate manipulation of a culture a threat to the very essence of man," asks the controversial Dr. B. F. Skinner of Harvard, "or, at the other extreme, an unfathomed source of strength for the culture that encourages it?" Skinner, who believes that man "is a psychological entity, and as such also largely man-made," long ago cast his vote for the control of behavior and the deliberate design of societies. Instead of just going along with whatever happens to happen to us, he is for "the effective use of man's intelligence in the construction of his own future."*

Characterizing governments and most social and political institutions as "behavioral technologies," Skinner calls for more imaginative "cultural inventors." Suppose he is right: even so, isn't it going against nature to control people by artificial means? Many people do argue that social manipulation is unnatural and should be avoided. But what is natural about the

---

* True enough, when Skinner uses his own intelligence to construct an imaginary future—as he did in his novel *Walden Two,* for example— people, including other scientists, begin to worry. Nevertheless, Skinner's utopian novel has influenced many people, and has even inspired the setting up of experimental communities patterned after Walden Two.

way we live? We spend our lives interfering with nature for our own benefit. We cultivate and cook food, we wear clothes, we build houses and jails and museums, we chlorinate water and pasteurize milk, we make fishhooks and automobiles and nuclear submarines. We go to school and to church and to war, and we pay taxes—in one form or another—to support them all. Hardly anything we do is strictly "according to nature."

Further, even a cursory look at our society—or any society, no matter how "primitive"—makes it clear that man's behavior is already controlled—by laws and customs, by social approval or disapproval, by educational and economic systems, by religious taboos and moral imperatives of every sort. Even in a free, democratic, nontotalitarian society, the behavior of every one of us is circumscribed in some way, in nearly every facet of our lives.

It is almost academic to ask whether or not we should control society. We do it now. But we do it badly. We need to do it better. As matters now stand, we passively let society happen to us, or we jerry-build it—a bit anxiously and a bit absent-mindedly, a piece at a time, a patch or a bung here and there to deal with a rip or a leak quickly sprung—with all kinds of painful and irrational results. Why not plan society more sensibly?

One of the reasons why not, of course, is that something about the idea alarms us. It smacks of totalitarianism, regimentation, and curtailment of individual freedoms, and though we may not be too happy with the results of our own drifting, we have been frightened by some of the large-scale attempts at social planning we have seen in our times. How do we devise social organizations—local, national, and international—that are efficient, rational, and stable, yet flexible enough to allow change and growth? How do we provide an overall cultural milieu that affords opportunities for the kind of fruitful and joyous work and play, and the flowering of individual human personality and creativity implicit in the concept of Genesis II? There is always the question, the same one our founding fathers had to face, of how you control the controllers, what mechanisms you build into your plans to prevent or punish the abuse of power. What we need to do is somehow *regulate* society without oppressive control of the individual.

One thing is certain: We do not have too much time to dawdle. The bulk of the job ought to be done while the controls at hand are still only social and social-scientific, before the large-scale advent of cheap and easy psychochemicals, streamlined ESB and people-farming techniques, and a really massive take-over by electronics and interlocking computer systems. To accept the inevitability of change does not mean that we have to accept all changes indiscriminately. We can and must still be selective. Nor, to accommodate progress, should we be required to tear down the old City of Man. It may be that a radical form of urban renewal will suffice. We could perhaps affect surprisingly sweeping changes within the pliable framework of our existing government, if there were proper guidance from the new breed of "cultural inventors" that Skinner hopes to see come forth.

When Einstein conceived the theory of relativity, certainly a wildly improbable sounding set of concepts, he looked around and found that others before him had invented all sorts of new mathematics—some of which had no previous use at all or had been devised for other purposes. He took what he needed and discarded the rest. In the same way social and political theorists may find what they need in surprising places.

The role of science itself will be a major consideration. As a general principle we will want to encourage research. We do not begin to know all that we need to know, and we cannot foresee the time when we will. Yet because we do not have the resources, we cannot give support to *every* kind of research. So we will have to think out our needs and our goals and establish the necessary priorities.

In the area of biological controls people do not want to be "improved" in the same ways, nor should they be coerced into accepting a standardized set of either characters or characteristics. Yet most of us would be willing to assign high priority to the need for making man better. We would agree almost unanimously that we could not too soon find ways to improve our health, our intelligence, and our sensory capacities; to live longer and more joyfully and more wisely. Where we must proceed with greater care, however, is in those areas where we feel least confident of what we want to attain and to become. We should be especially wary about the widespread application of principles or circumstances that would replace a vital and colorful diversity with a palling uniformity. Evolution, whether purposeful or haphazard, can produce monsters, creatures incapable of evolving further, or sickly specimens destined for extinction. With all his shortcomings man is so far the most highly advanced organism that we have any direct evidence of, and nature seems to be uncommitted as to whether his development is a progression, a regression, or an extinction. Thus, while actively seeking all means to make men wiser, it might be a good temporary decision to postpone the larger evolutionary decisions—since they could irrevocably close off other choices—until those wiser men are on the scene to make them. If world conditions force us to forgo the luxury of this waiting period, we can at least do our best to insure that biological controls are never instituted or exercised by minds whose primitive brain centers are in control of *them*. In short every item of knowledge must be explored to see what good it can do us and meticulously tested for possible harmful side effects. Where we have a choice, new disruptive elements should not be imposed upon a society already disrupted.

287

Many people believe we should seriously consider instituting much tighter controls over all research. In a recent *Saturday Review* article, for instance, Wilbur H. Ferry, a vice-president of the Fund for the Republic, expressed great qualms about permitting science and technology to continue their headlong progress unbraked. He proposed redrafting the Constitution, if necessary, to keep research in check—to guarantee that it proceeds at something less than a runaway pace and that its fruits are judiciously applied, with the welfare of the people paramount. This is an interesting idea. But he then suggested: "The sovereignty of the people must be reestablished, rules must be written and regulations imposed. The writing must be done by statesmen and philosophers consciously intent on the general welfare, *with the engineers and researchers [scientists] summoned from their caves to help in the doing when they are needed"* (italics mine). Such an attitude is at once unsound and dangerous.

It might be equally dangerous, of course, to take the opposite viewpoint: to assume naïvely, as many laymen seem to, that a scientist, because of his superior technical knowledge, can necessarily translate either his know-how or his objectivity to areas outside his specialty. A scientist can certainly fail to be objective—even about his specialty. He is as human as the next fellow and can just as easily be carried away by his emotions and prejudices. So we would be ill-advised to sit back and simply let the scientists run things. However, very few of them want to, so this does not pose a problem. When all disclaimers have been made, scientists *are* better trained as objective truth-seekers than most of the rest of us, and not all of them are incapable of applying their scientific objectivity to other areas of thought. The fact is that scientists have been more concerned than anyone about the implications of their discoveries. They are the ones who have been shouting the warning signals, trying—not always too successfully—to get the statesmen and philosophers, in whom Ferry seems to place such faith, to pay some attention to the matters that need attention.

One of the more reassuring aspects of scientific truth is that it is always tentative, always subject to change with the arrival of better data, and thus less likely to impose itself as absolute dogma. Not that scientists accept changes in their most cherished beliefs without resistance. Far from it. But here the process of change is much less painful than elsewhere, since change is one of science's own primary rules. A scientific law is accepted as true and acted upon as if it were true (because it does work, even if it works for the wrong reasons), with the full knowledge that the ultimate "truth" is still evolving. New insights, new facts, irrefutable data from experiments, cause the old truth to be discarded for the new. One cheering thing about all this is that, as a rule, new truths in science do not so much elbow out the old as incorporate them into a new context. The theory of relativity, while it is often said to have "toppled" the structure of Newtonian physics, actually left us Newtonian physics as a most valuable tool that we continue to use, as if it were true, because it is a close enough approximation to the truth for nearly all our earthly purposes.

It has been fashionable to denounce science for its dehumanizing influence. But as Dr. Bronowski points out in *Science and Human Values,* "like the other creative activities which grew from the Renaissance, science has humanized our values. Men have asked for freedom, justice, and respect precisely as the scientific spirit has spread among them. The dilemma of today is not that the human values cannot control a mechanical science. It is the other way about: The scientific spirit is more human than the machinery of governments. We have not let either the tolerance or the empiricism of science enter the parochial rules by which we still try to prescribe the behavior of nations." Or, as C. P. Snow has put it, scientists "have the future in their bones." To rule them out of the councils of power, to discriminate against them simply because they are scientists, would make as much sense as to deliberately cut out or lop off a vital organ.

It is obvious that we cannot arbitrarily assign any such passive role to the scientist. What should our attitude toward him be, then, when he speaks in a field outside his professional competence? The same as toward any other man: Listen to what he has to say, evaluate his thoughts on their merit. His scientific credentials should—in such a case—be neither a qualification nor a disqualification.

Even in the areas of human values, where theologians often flounder, scientists can offer some substantial contributions based on pioneering insights. For scientists do not want to do away with religion; rather they hope to find new foundations for it, to build a viable religion—what Sir Julian Huxley calls "religion without revelation"—whose truths, like those of science, will have high enough probability value for us to act on them as if they were true, but adjustable with minimal trauma to meet new knowledge and insights. It is held that such a religion—not requiring total allegiance or eternal vows of its adherents—would have a much greater chance of universal acceptance than any existing dogmatic religion, however modified, and however ecumenical the theological *Zeitgeist*.*

In the spelling out of his ideas on Evolutionary Humanism,

---

* No such science-oriented "religion" has yet been elucidated in thorough detail. But what might be called a religious movement has been evolving organically in the writings of a number of seminal thinkers, ranging all the way from the too-little-appreciated British philosopher Lancelot Law Whyte, whose essential book is *The Next Development in Man,* to the now celebrated ideas of the late French Jesuit philosopher-scientist Père Teilhard de Chardin, whose basic work is *The Phenomenon of Man.* The movement is called Evolutionary Humanism, and no one has assembled and articulated its ideas more elegantly or more eloquently than Huxley in a number of lectures, papers, articles, and books.

Sir Julian has emphasized again and again that the plans must be concerned with the individual. "The well-developed, well-patterned individual human being," he says in *The Humanist Frame,* "is, in a strictly scientific sense, the highest phenomenon of which we have any knowledge; and the variety of individual personalities is the world's highest richness." Huxley believes that when a man develops his individuality to the fullest degree, "in his own person he is realizing an important quantum of evolutionary possibility; he is contributing his own personal quality to the fulfillment of human destiny; and he has assurance of his own significance in the vaster and more enduring whole of which he is a part."

Sir Julian certainly does not argue that science should be enthroned above all other human activities. "The central belief of Evolutionary Humanism is that existence can be improved, that vast untapped possibilities can be increasingly realized, that greater fulfillment can replace frustration. This belief is now firmly grounded in knowledge: it could become in turn the firm ground for action."

So it is the scientists (among them, the school of humanistic psychology led by yea-sayers Dr. Abraham H. Maslow of Brandeis University and Dr. Carl Rogers of the Western Behavioral Sciences Institute) who are putting forth a vital effort to repair the damaged image of man. Everyone is invited to pitch in, especially the solitary thinker and creative artist: philosopher, social theorist, composer, graphic artist, filmmaker, and most particularly the literary artist—essayist, poet, dramatist, and novelist.

The arts hold up a mirror to man while probing deep beneath the surface to reveal the terrifying innards of man's soul. The mirror accommodates the anti-hero, the theater of protest, the questioning of current values, the exposure of hypocrisy and complacency and corruption in so many facets of contemporary life. But this is not all that art is made of, nor is it all that life is made of. Because of its one-sidedness, the image of man reflected in the works of too many contemporary creative artists is untrue because only *half* true.

In practice art does more than merely mirror what exists, what is happening. Out of the artist's reports, out of his imaginings, emerge powerful forces for influencing and determining *what happens next.* The artist helps create man's image of himself. In brief, art has consequences, just as science does. The scientist has finally recognized that he can no longer altogether disclaim responsibility for the consequences of his work. But the artists, especially those currently in vogue, seem unhampered by such restraint, and still pound home, with a maniacal sort of glee, their raucous messages of hopelessness.

If you tell a man long enough and loud enough that he is vile and vicious, he will begin to believe you and act accordingly. Man's image of himself today, as Barbara Tuchman has pointed out, has been horribly scarred, "with the result that man, at this moment in history, may no longer believe in his capacity to be good." Artist Earl Hubbard insists that artists and writers are duty-bound to create a new image that people—and especially young people—can aspire to, an image that will encompass the new dimensions man will be capable of taking on with his expanding powers. An imperative for artists is found in Malraux's commission to "fashion images of ourselves sufficiently powerful to deny our nothingness."

To create these images, our artistic creators must accept and deal with science and its implications. For the quality of our lives from now on will depend on science and the uses we make of it; its advance will alter our concepts of life, death, individual identity, and every variety of human relationship. Therefore, when I urge writers to deal with science, I am urging them to write their novels, their stories, their scenarios, their poems, and their plays—and whatever new forms they may evolve—with a deep, abiding awareness of human lives being lived in this new context.

In any planning of society, the structure and function of educational institutions (with education soon to encompass a lifetime) will be at the heart of it; and we are less likely to go wrong in our choices if we keep in mind what it is all to be designed for: the whole human being and his fulfillment in a regulated but free society. The educational establishment's major challenge will be to turn out people of high quality: people capable of constantly improving the quality of their own lives and interested in improving the lives of others; people who possess the necessary technical know-how, intellectual prowess, sensory awareness, personal and social responsibility to face cheerfully the unending ambiguities of the new age; people who are incapable of bestiality toward their fellowmen, who have no use for personal power unless it offers an opportunity to enhance the quality of life on earth for all mankind.

Can we turn out such men and women, citizens and leaders, and turn them out in variety and in quantity? We must and we can, if we do not lose hope. We cannot be sure that hope alone will take us very far along the road from Here to There, but we *can* be sure that hopelessness, all by itself, will lead us to the quick dead end. To strike hope dead is to commit a cardinal sin against the future.

Consider the unbelievable changes that have taken place with such dizzying rapidity in only a very few years: the breaking down of far-flung empires and the springing up of new nations; the massive shifting of alliances, with yesterday's friends today's foes; ecumenism among the churches, and the questioning of theology's basic tenets by theologians; sweeping transformations in race relations and in our sexual morality; unheard-of restraint in the face of provocations that would once have led instantaneously to all-out war. Most of these changes, though generated by ourselves, have been the result of random forces, not under much control. Yet to continue to let things happen that way as we approach Genesis II may be to play dice with disaster. Nevertheless, the fact that such astounding metamorphoses can take place almost inadvertently and in such brief time spans, gives us hope that we can achieve similarly astounding changes on purpose.

There is nothing in my personal experience, nor in anything I have ever heard or read, which convinces me that men, when they have alternate choices, *prefer* to be evil. My intellect as well as my instincts lead me rather to the opposite conclusion: that men have a positive yearning to be good—to love and be loved rather than to hate and be hated, to be proud and not ashamed of their humanity. The kind of higher good that Genesis II requires lies latent in many men; it surfaces sporadically in others; in certain rare individuals it is uppermost at most times. Some men keep it covered up defensively lest its exposure render them somehow vulnerable. In order to avoid disappointment, they refuse to hope.

If some genius with people knew how to tap this good simultaneously and universally (just as Hitler's *evil* genius was able to crystallize anti-Semitic feelings in Germany) or at least to bring it forth in a crucially significant number of "seed people," Genesis II will have arrived ahead of schedule. Ironically, the two men who seemed closest to possessing this kind of genius in our own country in our own time—John F. Kennedy and Martin Luther King—both fell victim to assassins' bullets. While waiting for their like to appear again, we must each carry a piece of the burden.

The "dust" from which man was created is believed by most scientists (and perhaps by most contempory theologians as well) to be metaphorical. Scientists believe—a tentative belief, of course, based on experimental evidence supplemented by sound theorizing—that at some hard-to-pinpoint time in the earth's past, the substances from which life was destined to be formed were ready and waiting in the primordial earth-soup. All they needed was the final energizing stroke, perhaps a powerful burst of cosmic radiation, to knock them together to form the first self-replicating molecules. Once it happened, the growth of organic substance on earth was explosive, and Genesis I was irrevocably under way.

I have the strong feeling, a feeling shared by many, that just such ripe-and-ready conditions exist today in the hopes and the hungers and the goodness of individuals all over the world, waiting for the energizing strokes that will carry us over the crest into Genesis II. Nor do we have to wait, this time, for any random bursts from the cosmos.

There is no inevitability now about where man goes, or how. It is up to you and me to set the goals and map the routes, and then, with daring tempered by reason, to launch our hazardous mission toward a yet-to-be discovered destiny.

Bibliography

AMA NEWS, "MDs, Clergy Discuss Prolonging Life," May 9, 1966.

ABELSON, PHILIP H., "Who Shall Live?" *Medical Science*, Nov., 1967.

ALDRICH, ROBERT A., "Reproduction and Gestation: An Assessment of Present-day Knowledge," an unpublished paper.

ALEXANDER, SHANA, "They Decide Who Lives, Who Dies," *Life*, Nov. 9, 1962.

ANDERSON, KENNETH N., "Can Man Be Modified to Live in Space?" *Today's Health*, Nov., 1963.

ASHLEY MONTAGU, M. F. (Ed.), *Marriage: Past and Present. A Debate Between Robert Briffault and Bronislaw Malinowski*, Porter Sargent, Boston, 1956.

ASIMOV, ISAAC, *The New Intelligent Man's Guide to Science*, Basic Books, New York, 1965.

———, "That Odd Chemical Complex, The Human Mind," *The New York Times Magazine*, July 3, 1966.

———, "Pills to Help Us Remember?" *The New York Times Magazine*, Oct. 9, 1966.

AUGER, PIERRE, "Limits to Science," *Bulletin of the Atomic Scientists*, Nov., 1965.

AVERY, T. L., C. L. COLE, E. F. GRAHAM, M. L. FAHNUNG, and V. G. PURSEL, "Investigations Associated with the Transplantation of Bovine Ova," in four parts, *Journal of Reproductive Fertility*, 1962.

AYD, FRANK J., JR., "The Hopeless Case," *Journal of the AMA*, Sept. 29, 1962.

BALFOUR-LYNN, STANLEY, "Parthenogenesis in Human Beings," *The Lancet*, June 30, 1956.

BEADLE, GEORGE W., "The New Genetics," *Brittanica Book of the Year*, 1964.

——— and MURIEL BEADLE, *The Language of Life*, Doubleday, New York, 1966.

BENJAMIN, HARRY, *The Transsexual Phenomenon*, The Julian Press, Inc., New York, 1966.

Bibliography

BELL, ROBERT R., "Some Emerging Sexual Expectations Among Women," paper given at AMA meeting, June 1967.

BERNARD, CLAUDE, *An Introduction to the Study of Experimental Medicine,* Collier Books, New York, 1961.

BHATTACHARYA, B. C., "Pre-arranging the sex of offspring," *New Scientist,* Oct. 15, 1964.

BLOOM, MURRAY T., "Explorer of the Human Brain," *Reader's Digest,* July 1958.

BONNER, JAMES, *The Molecular Biology of Development,* Oxford U. Press, New York and Oxford, 1965.

BOULDING, KENNETH, *The Meaning of the Twentieth Century,* Harper & Row, New York, 1964.

BRADBURY, RAY, "A Serious Search for Weird Worlds," *Life,* Oct. 24, 1960.

BRECHER, RUTH and EDWARD, "Every Sixth Teen-Age Girl in Connecticut—," *The New York Times Magazine,* May 29, 1966.

BRENTON, MYRON, *The American Male,* Coward McCann, New York, 1966.

BRILL, HENRY, "Contributions of Biological Treatment to Psychiatry," in Rinkel (below).

BRONSTED, H. V., "Warning and Promise of Experimental Embryology," *Bulletin of the Atomic Scientists,* March 1956.

BRONOWSKI, JACOB, *Science and Human Values,* Harper Torchbooks, New York, 1959.

——, *The Identity of Man,* The Natural History Press, New York, 1965.

BROWN, HARRISON, *The Challenge of Man's Future,* Viking Press, New York, 1954.

BUDRYS, A. J., "Mind Control is ☐ Good ☐ Bad," *Esquire,* May 1966.

BUNGE, R. G., "Further Observations on Freezing Human Spermatozoa," *The Journal of Urology,* Feb. 1960.

Bibliography

CALVIN, MELVIN, "The Origin of Life on Earth and Elsewhere," *Perspectives in Biology and Medicine,* Summer 1962.

CAMPBELL, ROBERT and NANCY GENET, "The Virus Enemy," *Life,* Feb. 16, 1966.

CANT, GILBERT, "An Overall View" in *Chemical Concepts of Psychosis,* (Max Rinkel, Ed.), McDowell–Obolensky, New York, 1958.

CARLOVA, JOHN, "When is a Patient Officially Dead?" *RISS* (Magazine for Residents, Interns and Senior Students, Oradell, N.J.), May 1964.

CHISHOLM, BROCK, "Future of the Mind," in Wolstenholme (below).

CLARK, LE MON and ISADORE RUBIN, "Dare I Have a Test-Tube Baby?" *Sexology,* Jan. 1961.

CLARKE, ARTHUR C., *Profiles of the Future,* Harper & Row, New York, 1962.

COHN, ROY, "Transplantation in Humans," *Stanford Today,* Winter 1967.

COMFORT, ALEX, *Ageing: The Biology of Senescence,* Holt, Rinehart & Winston, New York, revised 1964.

———, "Longevity of man and his tissues," in Wolstenholme (below).

———, *The Nature of Human Nature,* Harper & Row, New York, 1966.

———, "The Prevention of Ageing in Cells," *The Lancet,* Dec. 17, 1966.

COMMONER, BARRY, "The Elusive Code of Life," *Saturday Review,* Oct. 1, 1966.

———, "The Implications of Molecular Biology for Man," paper given at the New School for Social Research, Apr. 21, 1967.

COUGHLAN, ROBERT, "Control of the Brain," a two-part article, *Life,* Mar. 8 and Mar. 15, 1963.

CREECH, OSCAR, "The Shape of Medicine in 1990," *Medical World News,* Nov. 25, 1966.

CROW, JAMES F., "Mechanisms and Trends in Human Evolution," *Daedalus,* Summer 1961.

DAVIDSON, BILL, "Probing the Secret of Life," *Collier's,* May 14, 1954.

Bibliography

DE BAKEY, MICHAEL E., "Prospects for and Implications of the Artificial Heart," *Journal of Rehabilitation*, March–April 1966.

DELGADO, JOSE M. R., "Evolution of Physical Control of the Brain," the James Arthur Lecture at the American Museum of Modern History, 1965.

DE MARS, ROBERT, "Investigations in Human Genetics with Cultivated Human Cells: A Summary of Present Knowledge," in Sonneborn (below).

DE MARTINO, MANFRED F. (Ed.), *Sexual Behavior and Personality Characteristics*, Grove Press, New York, 1966.

DEMIKHOV, V. P., *Experimental Transplantation of Vital Organs*, Consultants Bureau, New York, 1962.

DIAMOND, EDWIN, "Are We Ready to Leave Our Bodies to the Next Generation?" *The New York Times Magazine*, Apr. 21, 1968.

DOBZHANSKY, THEODOSIUS, *Mankind Evolving*, Yale U. Press, New Haven, 1962.

——, *Heredity and the Nature of Man*, Harcourt, Brace & World, New York, 1964.

DOWLING, HARRY F., "Human Experimentation in Infectious Diseases," *Journal of the AMA*, Nov. 28, 1966.

Drug Abuse: Escape to Nowhere, Smith Kline & French Labs, Philadelphia, 1967.

DUBOS, RENÉ, *Mirage of Health*, Harper & Bros., New York, 1959.

——, *Man Adapting*, Yale U. Press, New Haven, 1965.

——, "Humanistic Biology," *American Scientist, 53*, 1965.

DURY, DAVID, "Sex Goes Public: A Talk with Henry Miller," *Esquire*, May 1966.

ECCLES, J. C., *The Neurophysiological Basis of Mind*, Oxford U. Press, 1953.

EDWARDS, ROBERT, and RICHARD GARDNER, "Choosing Sex Before Birth," *New Scientist*, May 2, 1968.

302

Bibliography

EISENBERG, LUCY, "Genetics and the Survival of the Unfit," *Harper's,* Feb. 1966.

ELKES, JOEL J., "Psychotropic Drugs: Observations on Current Views and Future Problems," in *Lectures on Experimental Psychiatry,* Pittsburgh U. Press, 1961.

——, "Some Prospects in the Neurological Sciences," Deerfield Foundation Lecture at Gallery of Modern Art, New York, Dec. 5, 1965.

ELKINTON, J. RUSSELL, "Moral Problems in the Use of Borrowed Organs, Artificial and Transplanted," *Annals of Internal Medicine,* Feb. 20, 1964.

——, "Medicine and the Quality of Life," *Annals of Internal Medicine,* March 1966.

——, "When Do We Let the Patient Die?" *Annals of Internal Medicine,* March 1968.

ENGEL, LEONARD, *The New Genetics,* Doubleday, New York, 1967.

ETTINGER, ROBERT C. W., *The Prospect of Immortality,* Doubleday, New York, 1964.

——, "Cryonics and the Purpose of Life," *Christian Century,* Oct. 4, 1967.

FALLACI, ORIANA, "The Dead Body and the Living Brain," *Look,* Nov. 28, 1967.

FARSON, RICHARD E. (Ed.), *Science and Human Affairs,* Science and Behavior Books, Palo Alto, 1965.

FERRY, WILBUR H., "Must We Rewrite the Constitution to Control Technology?" *Saturday Review,* Mar. 2, 1968.

FISHER, ALAN E., "Chemical Stimulation of the Brain," *Scientific American,* June 1964.

FLETCHER, JOSEPH, "Methods of Fertility Control: A Moral Tension," and "Death and Medical Initiative," *Lectures on Medical Ethics 1963-1964* (mimeographed), Yale U. School of Medicine.

——, *Situation Ethics: The New Morality,* The Westminster Press, Philadelphia, 1966.

Bibliography

FREEDMAN, TOBEY, *Man in Space*, North American Aviation, Inc. booklet.

GAGNON, JOHN H., "Sexuality and Sexual Learning in the Child," *Psychiatry*, Aug. 1965.

GALTON, LAWRENCE, "One of the Great Mystery Stories of Medicine," *The New York Times Magazine*, Nov. 6, 1966.

GAZZANIGA, MICHAEL S., "The Split Brain in Man," *Scientific American*, Aug. 1967.

GERARD, RALPH W., "What is Memory?" *Scientific American*, Sept. 1953.

GILMAN, WILLIAM, "Tampering with Our Genetic Blueprint," *The New Republic*, Nov. 22, 1966.

GLASS, H. BENTLEY, *Science and Ethical Values*, University of North Carolina Press, Chapel Hill, 1965.

GOLDSTEIN, PHILIP, *Triumphs of Biology*, Doubleday, New York, 1965.

GOODMAN, PAUL, *Growing Up Absurd*, Random House, New York, 1960.

GORER, GEOFFREY, "Man Has No 'Killer' Instinct," *The New York Times Magazine*, Nov. 27, 1966.

GRUNFELD, FREDERIC V., "Eine Kleine Carbon Copy," *The Reporter*, Jan. 26, 1967.

GUTTMACHER, ALAN F., "The Role of Artificial Insemination in the Treatment of Sterility," *Obstetrical and Gynecological Survey*, Dec. 1960.

HAFEZ, E. S. E. (Ed.), *Reproduction in Farm Animals* (2nd ed.), Lea & Febiger, Philadelphia, 1968.

HALACY, D. S., JR., *Cyborg: Evolution of the Superman*, Harper & Row, New York, 1965.

HALDANE, J. B. S., "Biological Possibilities for the human species in the next ten thousand years," in Wolstenholme (below).

HAMILTON, MICHAEL, "Moral Issues in Medical Technology," sermon delivered at the Washington Cathedral, Nov. 20, 1966.

HAMLIN, HANNIBAL, "Life or Death by EEG," *Journal of the AMA*, Oct. 12, 1964.

Bibliography

————, "The Moment of Death," *Medical Tribune*, May 12, 1965.

HANEY, BETTE M. and M. W. OLSEN, "Parthenogenesis in premature and newly laid turkey eggs," *Journal of Experimental Zoology*, Dec. 1958.

HARRIS, MORGAN, *Cell Culture and Somatic Variation*, Holt, Rinehart & Winston, New York, 1964.

HEATH, ROBERT G., "Electrical Self-Stimulation of the Brain in Man," *American Journal of Psychiatry*, Dec. 1963.

————, "Developments Toward New Physiologic Treatments in Psychiatry," *Journal of Neuropsychiatry*, Vol. 5, No. 6, 1964.

————, "Schizophrenia: Biochemical and Physiologic Aberrations," *International Journal of Neuropsychiatry*, Vol. 2, No. 6, 1966.

————, "Schizophrenia: Pathogenetic Theories," *International Journal of Psychiatry*, May 1967.

———— (Ed.), *The Role of Pleasure in Behavior*, Harper & Row, New York, 1964.

HOAGLAND, HUDSON, "Potentialities in the control of behavior," in Wolstenholme (below).

HOFFER, ERIC, "A Strategy for the War with Nature," *Saturday Review*, Feb. 15, 1966.

HOWARD, JANE, "Inhibitions Thrown to the Gentle Winds," *Life*, July 12, 1968.

HUXLEY, ALDOUS, *The Doors of Perception*, Harper, New York, 1964.

HUXLEY, JULIAN, *Religion Without Revelation*, New American Library, New York, 1958.

————, "The Future of Man—Evolutionary Prospects," in Wolstenholme (below).

————, "Eugenics in Evolutionary Perspective," *Perspectives in Biology and Medicine*, Winter 1963.

———— (Ed.), *The Humanist Frame*, Harper & Row, New York, 1961.

HYDÉN, HOLGER, "Satellite Cells in the Nervous System," *Scientific American*, December 1961.

Bibliography

JACKSON, BRUCE, "White-Collar Pill Party," *Atlantic,* August 1966.

KANTER, SHAMAI, "If Man Creates Life He Is Still Man," *National Jewish Monthly,* Nov. 1963.

KATZ, JAY, "Experiments on People—What Are the Limits?" lecture given at Yale U. School of Medicine, Jan. 14, 1965.

KETTLE, JOHN, series on the year 2000 in *Monetary Times,* 1967.

KEVORKIAN, JACK, "The Nobler Execution," *Ararat,* Summer 1961.

———, "Capital Punishment or Capital Gain?" in *Crime in America* (Herbert A. Bloch, Ed.), Philosophical Library, New York.

KINZEL, AUGUSTUS B., "Engineering, Civilization and Society," *Science,* June 9, 1967.

KLEIN, ISAAC, "Autopsy and Abortion," and "Sterilization, Contraception and Artificial Insemination," *Lectures on Medical Ethics 1963–1964* (mimeographed), Yale U. School of Medicine.

KLINE, NATHAN S., "Psychopharmaceuticals: Effects and Side Effects," *Bulletin of the WHO,* Vol. 21, 1959.

———, "Comprehensive Therapy of Depressions," *Journal of Neuropsychiatry,* Feb. 1961.

———, "The Alteration of 'Natural' Biological States by LSD," *The Hastings Law Journal,* March 1966.

——— and Manfred Clynes, "Cyborgs and Space," *Astronautics,* Sept. 1960.

———, "Drugs, Space, and Cybernetics: Evolution Cyborgs" in *Psychophysiological Aspects of Space Flight, Columbia U. Press,* New York, 1961.

KOESTLER, ARTHUR, *The Ghost in the Machine,* Macmillan, New York, 1967.

KOPROWSKI, HILARY, "Future of infectious and malignant diseases," in Wolstenholme (below).

KRECH, DAVID, "The Chemistry of Learning," *Saturday Review,* Jan. 20, 1968.

Bibliography

LANDERS, RICHARD R., *Man's Place in the Dybosphere*, Prentice-Hall, Englewood Cliffs, 1966.

LEAR, JOHN, "Who Should Govern Medicine?" *Saturday Review*, June 5, 1965.

———, "Do We Need Rules for Experiments on People?" *Saturday Review*, Feb. 5, 1966.

———, "Policing the Consequences of Science," *Saturday Review*, Dec. 2, 1967.

———, "Transplanting the Heart," *Saturday Review*, Jan. 6, 1968.

———, "A Realistic Look at Heart Transplants," *Saturday Review*, Feb. 3, 1968.

LEDERBERG, JOSHUA, "Biological future of man," in Wolstenholme (below).

———, "Experimental Genetics and Human Evolution," *Bulletin of the Atomic Scientists*, Oct. 1966.

———, "Biomedical Research: Its Side-effects and Challenges," *Stanford MD*, Oct. 1967.

LESSING, LAWRENCE and the EDITORS of *Fortune*, *DNA: At the Core of Life Itself*, Macmillan, New York, 1967. (Originally a four-part series in *Fortune*.)

LEVINE, SEYMOUR, "Sex Differences in the Brain," *Scientific American*, Apr. 1966.

Life, "A Turkey That Never Had a Father," Apr. 10, 1956.

———, " 'Code 99!' Calls a Hospital to Battle Against Death," Feb. 28, 1964.

———, "A Living Brain," July 31, 1964.

———, "The Fantastic Drug That Creates Quintuplets," Aug. 13, 1965.

———, "The Control of Life," a four-part series, Sept. 10, Sept. 17, Sept. 23, and Oct. 1, 1965.

———, "Max the Lifesaver," Jan. 28, 1966.

Bibliography

——, "A Remarkable Mind Drug Suddenly Spells Danger: LSD," Mar. 25, 1966.

——, "A Patient's Gift to the Future of Heart Repair," May 6, 1966.

——, "Two Sex Researchers on the Firing Line," June 24, 1966.

——, "Womb with Windows," July 29, 1966.

——, "Marijuana: Millions of Turned-on Users," July 7, 1967.

——, "Deathproof Patient for Student Doctors," Dec. 8, 1967.

——, "Gift of a Heart," Dec. 15, 1967.

——, "The No-Nonsense Heart Man of Houston," August 2, 1968.

LOMAX, LOUIS E., "Sperm Bank—The Brutal Truth We Dare Not Face," *Pageant,* Jan. 1958.

LONGMORE, DONALD, "Implants or Transplants?" *Science Journal,* February 1968.

LURIA, SALVADOR E., "Directed Genetic Change: Perspectives from Molecular Genetics," in Sonneborn (below).

MD, "Reconstituted Organs," Nov. 1960.

MAGOUN, H. W., "Neural Plasticity and the Memory Process," in Rinkel (below).

MAISEL, ALBERT Q., "Can Science Prolong Our Useful Years?" *Reader's Digest,* Jan. 1962.

MAKARENKO, A. S., *The Collective Family: A Handbook for Russian Parents,* Doubleday Anchor Books, New York, 1967.

MANN, THOMAS, "Dostoevsky—in Moderation," a preface to *The Short Novels of Dostoevsky,* Dial Press, New York, 1945.

MARTIN, DONALD C., MILTON RUBINI and VICTOR J. ROSEN, "Cadaveric Renal Homotransplantation with Inadvertent Transplantation of Carcinoma," *Journal of the AMA,* May 31, 1965.

MASLOW, ABRAHAM H., *Toward a Psychology of Being,* D. Van Nostrand, Princeton, 1962.

——, "Love in Self-Actualizing People," in DeMartino (above).

Bibliography

MASSACHUSETTS GENERAL HOSPITAL, *Human Studies—Guiding Principles and Procedures* (booklet), Boston, 1967.

MASTERS, WILLIAM H. and VIRGINIA E. JOHNSON, *Human Sexual Response*, Little, Brown, Boston, 1966.

———, Interview in *Playboy*, May 1968.

———, "The Morality of Sex Research," paper presented at meeting of Academy of Religion and Mental Health, New York, Apr. 30, 1968.

MCINTOSH, DUNCAN A. (Capt., USAF-MC) and colleagues, Lackland AFB, Texas, "Homotransplantation of a Cadaver Neoplasm and a Renal Homograft," *Journal of the AMA,* June 28, 1965.

MCWHIRTER, WILLIAM A., " 'The Arrangement' at College," *Life,* May 31, 1968.

MEAD, MARGARET, *Male and Female,* William Morrow, New York, 1949.

MEDAWAR, PETER B., *The Future of Man,* New American Library, New York, 1961.

———, "The Great Problems: A Program for the Natural Sciences," address at Cornell U. Charter Week Symposium, Apr. 28, 1965.

Medical Tribune, "Artificial Insemination Case Makes 'Medicological History,' " Sept. 20, 1963.

———, "Organ Transplants Pose Moral Issues," Apr. 25, 1964.

———, "Many Cardiologists Voice Wariness on Transplants," Jan. 29, 1968.

Medical World News, "Isolated Brain Lives 7 Hours," Oct. 25, 1963.

———, "Ca Researchers Face $1-Million Suit," June 19, 1964.

———, "Human Ova Fertilized After Frozen Storage," Mar. 5, 1965.

———, "Sweden Weighs Trial of MDs Over Consent to Transplants," Apr. 2, 1965.

———, "Unethical Testing Under Fire," Apr. 9, 1965.

———, "Updating the Definition of Death," Apr. 28, 1967.

———, "Heart Transplants: How Many, How Soon?" Feb. 16, 1968.

Bibliography

MICHELMORE, SUSAN, *Sexual Reproduction,* The Natural History Press, Garden City, 1964.

MILLER, NEAL E., "Physiological and Cultural Determinants of Behavior," in *The Scientific Endeavor* (below).

MINSKY, MARVIN L., "Artificial Intelligence," *Scientific American,* Sept. 1966.

MONEY, JOHN, "Phantom Orgasm in the Dreams of Paraplegic Men and Women," *Archives of General Psychiatry,* Oct. 1960.

—— (Ed.), *Sex Research: New Developments,* Holt, Rinehart & Winston, New York, 1965.

MOORE, FRANCIS D., *Give and Take: A History of Transplantation,* Doubleday, New York, 1964.

——, "Tissue Transplants," *The Nation,* Apr. 5, 1965.

MOORE, KEITH L., "The Vulnerable Embryo: Causes of Malformation in Man," *Manitoba Medical Review,* June–July 1963.

—— and JEAN C. HAY, "Human Chromosomes," a review article in two parts, *The Canadian Medical Association Journal,* May 18 and 25, 1963.

MUELLER, GERHARD O. W., "Toward Ending the Double-Standard of Sexual Morality," *Journal of Offender Therapy,* Vol. 8, No. 1, 1964.

MULLER, HERMANN J., "Should We Weaken or Strengthen Our Genetic Heritage?" *Daedalus,* Summer 1961.

——, "Genetic progress by voluntarily conducted germinal choice," in Wolstenholme (below).

——, "Means and Aims in Human Betterment," in Sonneborn (below).

NAISMITH, GRACE, *Private and Personal,* David McKay, New York, 1966.

NELSON, ROBERT F., *We Froze the First Man,* Dell Publishing Co., New York, 1968.

Newsweek, "LSD and the Drugs of the Mind," May 9, 1966.

——, "The Promise and Perils of Transplant Surgery," Dec. 18, 1967.

Bibliography

O'DONNELL, THOMAS J., "Treatment of the Terminally Ill," *Lectures on Medical Ethics 1963–1964* (mimeographed), Yale U. School of Medicine.

OLDS, JAMES, "Pleasure Centers in the Brain," *Scientific American,* Oct. 1956.

OLSEN, M. W., "Fatherless Turkey," *Journal of the American Veterinary Medical Association,* Jan. 1, 1959.

———, "Performance Record of a Parthenogenetic Turkey Male," *Science,* Dec. 2, 1960.

———, "Nine Years Summary of Parthenogenesis in Turkeys," *Proceedings of the Society for Experimental Biology and Medicine,* v. 105, 1960.

OPARIN, ALEKSANDR I., *The Origin of Life,* Dover Publications, New York, 1953.

Paese Sera, "Sensational Scientific Experiment at Bologna: Artificial Insemination Obtained for the First Time Outside the Human Body," Jan. 12, 1961.

PAGE, IRVINE H., "Death with Dignity," *Modern Medicine,* Oct. 15, 1962.

———, "Prolongation of Life in an Affluent Society," *Modern Medicine,* Oct. 14, 1963.

———, Artificial Prolongation of Life," *Modern Medicine,* Oct. 26, 1964.

———, "Medical Research As I See It," an unpublished paper.

PAPPWORTH, M. H., *Human Guinea Pigs,* Routledge & Kegan Paul, London, 1967.

PARKES, A. S., *Sex, Science and Society,* Oriel Press Ltd., Newcastle upon Tyne, 1966.

PARKS, JOHN, "Trends Toward Medical Practice in 1975," paper given at meeting of the American Association for the Advancement of Science, Washington, Dec. 28, 1966.

PETRUCCI, DANIELE, "Producing Transplantable Human Tissue in the Laboratory," *Discovery,* July 1961.

311

Bibliography

PICKERING, GEORGE, "Degenerative Diseases: Past, Present and Future," paper given at Columbia U.-Merck Sharp & Dohme symposium, New York, May 26, 1966.

PINCUS, GREGORY, *The Control of Fertility*, Academic Press, New York, 1965.

PINES, MAYA, "Exploring the Brain's Uncharted Realms," *The Reporter*, May 15, 1958.

PONTECORVO, GUIDO, "Prospects for Genetic Analysis in Man," in Sonneborn (below).

PUBLIC BROADCAST LABORATORY, script of broadcast on medical morality, presented on WNDT-TV, New York, Jan. 21, 1968.

REISS, IRA L., "How and Why America's Sex Standards Are Changing," *Trans-Action*, March 1968.

RINKEL, MAX (Ed.), *Biological Treatment of Mental Illness*, L. C. Page, New York, 1966.

ROBIN, EUGENE D., "Rapid Scientific Advances Bring New Ethical Questions," *Journal of the AMA*, Aug. 24, 1964.

——— and CARROLL E. CROSS, "Lung Transplantation—Past, Present and Future," *Annals of Internal Medicine*, Nov. 1966.

ROGERS, CARL R., "A Humanistic Conception of Man," in Farson (above).

ROSENFELD, ALBERT, "The Futuristic Riddle of Reproduction," *Coronet*, Feb. 1959.

———, "The Great New Dream of Dr. Salk," *Life*, Feb. 8, 1963.

———, "The Last Barrier is Man Himself," *Life*, Oct. 2, 1964.

———, "Drama of Life Before Birth," *Life*, Apr. 30, 1965.

———, "The New Man: What Will He Be Like?" *Life*, Oct. 1, 1965.

———, "A Laboratory Study of Sexual Behavior," *Life*, Apr. 22, 1966.

———, "Science is Where the Action Is," *Life*, July 29, 1966.

———, "10,000-to-1 Payoff," *Life*, Sept. 27, 1967.

———, "Drugs That Even Scare Hippies," *Life*, Oct. 27, 1967.

Bibliography

———, "Heart Transplant: Search for an Ethic," *Life*, Apr. 5, 1968.

———, "The Scientists' findings: more sex, less promiscuity," *Life*, May 31, 1968.

———, "The Psychobiology of Violence," *Life*, June 21, 1968.

——— and ALICIA HILLS, "DNA's Code: Key to All Life," *Life*, Oct. 4, 1963.

ROSTAND, JEAN, *Can Man Be Modified?* Basic Books, Inc., New York, 1959.

——— and ALBERT DELAUNAY (Eds.), *Man of Tomorrow*, Vol. 8 of the *Encyclopedia of the Life Sciences*, Doubleday, Garden City, 1966.

RUBIN, BERNARD, "Psychological Aspects of Human Artificial Insemination," *Archives of General Psychiatry*, August 1965.

SALK, JONAS E., "Biology in the Future," paper given at M.I.T. Centennial, Apr. 8, 1961.

———, "Human Purpose—A Necessity," commencement lecture at the Phillips Exeter Academy, June 11, 1961.

———, "Education—for What?" paper given at NEA convention, June 26, 1961.

SANDERS, MARION K., "The Sex Crusaders from Missouri," *Harper's*, May 1968.

SCHEINFELD, AMRAM, *Your Heredity and Environment*, J. B. Lippincott, Philadelphia, 1965.

SCHMECK, HAROLD M., JR., *The Semi-Artificial Man*, Walker & Co., New York, 1965.

SCHMITT, F. O. (Ed.), *Macromolecular Specificity and Biological Memory*, M.I.T. Press, Cambridge, 1962.

Science, "Human Experimentation: New York Verdict Affirms Patient's Rights," Feb. 11, 1966.

Science News Letter, "Is Martyrdom Ethical?" Apr. 3, 1965.

SELYE, HANS, *The Stress of Life*, McGraw-Hill, New York, 1956.

313

Bibliography

SHAPLEY, HARLOW, *The View from a Distant Star*, Dell Publishing Co., New York, 1964.

SHERMAN, JEROME K., "Improved Method for Frozen Storage of Human Spermatozoa," paper given at annual meeting of Federation of American Societies for Experimental Biology, 1962.

SHETTLES, LANDRUM B., "Parthenogenetic Cleavage of the Human Ovum," *Bulletin of the Sloane Hospital for Women*, June 1957.

SHKLOVSKII, I. S., and CARL SAGAN, *Intelligent Life in the Universe*, Holden-Day, San Francisco, 1966.

SHUMWAY, NORMAN E., "Cardiac Transplantation," *The Heart Bulletin*, May–June 1963.

SIMON, WILLIAM and JOHN H. GAGNON, "Towards a New Man in a New City," *The Intercollegian*, Spring 1967.

———, "Pornography: The Social Sources of Sexual Scripts," paper given at Society for the Study of Social Problems, San Francisco, August 1967.

———, "The Pedagogy of Sex," *Saturday Review*, Nov. 18, 1967.

SINSHEIMER, ROBERT, "The End of the Beginning," paper presented at Cal Tech 75th Anniversary celebration, Oct. 26, 1966.

SKINNER, B. F., *Walden Two*, Macmillan, New York, 1948.

———, "The Design of Cultures," *Daedalus*, Summer 1961.

SNOW, C. P., *The Two Cultures: and a Second Look*, Cambridge U. Press, London, 1965.

SOLOMON, DAVID (Ed.), *The Marihuana Papers*, Bobbs-Merrill, Indianapolis, 1966.

SONNEBORN, TRACY M. (Ed.), *The Control of Human Heredity and Evolution*, Macmillan, New York, 1965. (Originally a symposium at Ohio Wesleyan U.)

SPENCER, STEVEN M., "The Birth Control Revolution," *Saturday Evening Post*, Jan. 15, 1966.

———, "The Pill That Helps You Remember," *Saturday Evening Post*, Sept. 24, 1966.

314

Bibliography

SPERRY, R. W., "The Great Cerebral Commissure," *Scientific American,* Jan. 1964.

STAHL, FRANKLIN W., *The Mechanics of Inheritance,* Prentice-Hall, Englewood Cliffs, 1964.

STILL, JOSEPH W., "Levels of Human Death and Life, and Some Religious and Ethical Implications," paper given at annual meeting of Federation of American Societies for Experimental Biology, April 18, 1968.

SULLIVAN, WALTER, *We Are Not Alone: The Search for Intelligent Life on Other Worlds,* McGraw-Hill, New York, 1964.

SZENT-GYÖRGYI, ALBERT, "The promise of medical science," in Wolstenholme (below).

TATUM, EDWARD L., "Perspectives from Physiological Genetics," in Sonneborn (above).

———, "Genetic Determinants," in *The Scientific Endeavor* (below).

———, "Molecular Biology, Nucleic Acids, and the Future of Medicine," paper given at Columbia U.-Merck Sharp & Dohme symposium, New York, May 26, 1966.

TAYLOR, GORDON R., *The Biological Time Bomb,* Thames & Hudson, London, 1968.

TEILHARD DE CHARDIN, PIERRE, *The Phenomenon of Man,* Harper & Row, New York, 1961.

———, *The Future of Man,* Harper & Row, New York, 1964.

The Lancet, "Parthenogenesis in Mammals?" Nov. 5, 1955.

The Scientific Endeavor, Rockefeller University Press, New York, 1965. (Originally a symposium to celebrate the centennial of the National Academy of Sciences.)

Time, "Parthenogenesis?" Nov. 21, 1955.

———, "The Problem of Old Age," July 23, 1956.

———, "Sex to Order?" Oct. 7, 1957.

———, "The Secret of Life," July 14, 1958.

315

———, "Survival of the Unfit?" June 16, 1958.

———, "Spare Parts from Chimp to Man," Dec. 27, 1963.

———, "What Darwin Didn't Know," May 29, 1964.

———, "Sex by Sedimentation," July 17, 1964.

———, "Questions of the Heart," July 17, 1964.

———, "The Age of Alloplasty," Jan. 1, 1965.

———, "Transfusions in the Womb," Jan. 15, 1965.

———, "Formula of Fugu," Jan. 25, 1965.

———, "The Riddle of A.I.," Feb. 25, 1966.

———, "A Molecule for Memory?" Jan. 7, 1966.

———, "The Ultimate Operation," Dec. 15, 1967.

UNGAR, GEORGES, "Chemical transfer of learned behavior," paper given at AAAS meeting, New York, Dec. 29, 1967.

VALLEE, BERT L. and WARREN E. C. WACKER, "Medical Biology: A Perspective," *Journal of the AMA*, May 11, 1963.

VON HULST, ERICH and URSULA VON ST. PAUL, "Electrically Controlled Behavior," *Scientific American*, March 1962.

WALD, GEORGE, "The Origins of Life," in *The Scientific Endeavor* (above).

WARSHOFSKY, FRED, *The Rebuilt Man*, Thomas Y. Crowell, New York, 1965.

WHITMAN, ARDIS, "Is Marriage Still Sacred?" *Redbook*, Feb. 1967.

WHITNEY, LEON F., "The successful transfer of ovaries between dogs of different breed," *Science*, May 24, 1946.

———, "Ovarian Transplantation in Dogs," *Veterinary Medicine*, January 1947.

WHYTE, LANCELOT L., *The Next Development in Man*, Henry Holt, 1948.

WIENER, NORBERT, *The Human Use of Human Beings*, Doubleday Anchor Books, New York, 1954.

316

——, *Cybernetics* (2nd ed.), M.I.T. Press and John Wiley, New York, 1961.

WILLETT, E. L., "Egg Transfer and Superovulation in Farm Animals," *Iowa State Journal of Science,* Sept. 1953.

——, P. J. Buckner and G. L. Larson, "Three Successful Transplantations of Fertilized Bovine Eggs," *Journal of Dairy Science,* May 1953.

WILLIAMS, GLANVILLE, *The Sanctity of Life and the Criminal Law,* Alfred A. Knopf, New York, 1967.

WILLIAMSON, WILLIAM P., "Life or Death—Whose Decision?" *Journal of the AMA,* Sept. 5, 1966.

WILLIER, BENJAMIN H. and JANE M. OPPENHEIMER (Eds.), *Foundations of Experimental Embryology,* Prentice-Hall, Englewood Cliffs, 1964.

WINCHESTER, JAMES H., "Babies Without Fathers!" *Sunday Mirror Magazine,* Jan. 29, 1956.

WITSCHI, EMIL, "Sex Reversal in Animals and in Man," *American Scientist,* Sept. 1960.

WOLSTENHOLME, GORDON (Ed.), *Man and His Future,* Little Brown, Boston, 1963. (Originally a Ciba Foundation Symposium in London.)

WOODRUFF, M. F. A., "Ethical Problems in Organ Transplantation," *British Medical Journal,* June 6, 1964.

WOOLDRIDGE, DEAN E., *Mechanical Man: The Physical Basis of Intelligent Life,* McGraw-Hill, New York, 1968.

WOOLLCOTT, JOAN, "The Mortal Cell," *Medical Affairs,* Sept. 1967.

YOUNG, WARREN R., "It's a Miracle That We Save Any of Them," *Life,* Dec. 2, 1966.

Index

Index

Index

encounter groups, 163
Enders, Dr. John, 23
enzymes, 232–235
   and mental illness, 234–235
epilepsy, 205
Ervin, Dr. Frank R., 201
Ettinger, Robert C. W., 45–47, 55, 65–66, 71, 277
Etzioni, Dr. Amitai, 115
eugenics, 140
euphemics, 140
euthanasia, 79
Evolutionary Humanism, 290–291
experimental medicine
   legal rulings, 85–87
   when justified, 85–87, 95–96
   ethics of consent for, 86–87
   why necessary, 88–89
   conflict with medical ethics, 89–95
   Nazi research, 91–92
   guidelines for, 95–96

Fanny Hill, 167
Farenheit 451, 134
Farson, Dr. Richard, 11, 162–163
Ferry, Wilbur H., 288
Fletcher, Reverend Dr. Joseph, 79, 182, 184
Foerster, Dr. Heinz von, 255, 259
Forssman, Dr. Werner, 63, 83
Fox, Dr. Sidney W., 148
freezer program, 45–47, 55, 65–67, 69, 70–71, 277
Freud, Dr. Sigmund, 227, 283
Fuller, Lon, 88

Gagnon, Dr. John, 170, 172, 185
Gaillard, Dr. Peter, 24
Gardner, Richard, 138–139
Gaudart d'Allaines, François de, 60
genetic disorders, 138–139, 158, 228–229
genetics
   of parthenogenesis, 112–114
   synthesis of genes, 140
   see also DNA; chemistry, cerebral
genetic surgery, 126–128, 135, 151
generation gap, 164–166
Georgetown University Medical Center, 97
Gerard, Dr. Ralph W., 190, 228, 254
geriatrics, 32
gerontology, 32–33

Gey, Dr. George O., 23
Give and Take, 51
Glass, Dr. H. Bentley, 33, 140, 142
Goldfinger, 218
Goodlin, Dr. Robert, 118
Goodman, Paul, 165–166
Great, Rev. Kenneth, 171, 183
Grossman, Dr. S. P., 212
Gurdon, Dr. J. B., 127, 141
Guttmacher, Dr. Alan F., 151

Hafez, Dr. E. S. E., 120–126, 132, 159, 161
Haldane, Dr. J. B. S., 42, 89, 91, 111, 113, 130, 140, 142–143
Hall, Dr. C. William, 58
hallucinogens, 216, 230
Hamilton, Canon Michael, 48
Hamlin, Dr. Hannibal, 59, 60, 74
Handler, Dr. Phillip, 27–28
Hardy, Dr. James D., 59
Harman, Dr. Denham, 38–39
Harrad Experiment, The, 173
Harrison, Dr. Ross, 21
Hayflick, Dr. Leonard, 129
Heape, Dr. Walter, 121
heart massage, 43–44
heart transplants, *see* transplants, heart
Heath, Dr. Robert G., 196, 204–205, 236
Heimberger, Dr. Robert, 202
HeLa Strain, 23, 129
Heredity and the Nature of Man, 136
hermaphroditism, 110
Hess, Dr. Walter R., 195
heterosis, 158–159
Hoagland, Dr. Hudson, 245
Hobbes, Thomas, 283
Hoffer, Dr. Abraham, 236
Hoffer, Eric, 281
Howard, Sidney, 90
Hubbard, Earl, 292
Human Destiny, 147
Human Guinea Pigs, 86
Humanist Frame, The, 291
Humphrey, Hubert, 75
Hunter, Dr. G. L., 121
Huxley, Aldous, 120, 179, 221, 240
Huxley, Sir Julian, 4, 7, 146, 154, 190, 290
Huxley, T. H., 248
Hydén, Dr. Holger, 250, 253, 256

323

Index

Index

Index

reproduction, scientific control of
  genetic surgery, 126–128, 135, 140, 144
  medical intervention in conception, 160–161
  restricted for genetic reasons, 136–139
  see also: abortion; artificial inovulation; artificial insemination; hermaphroditism; *in vitro* embryology; parthenogenesis; superovulation
Rh-negative transfusions, 118
Rhoads, Dr. Paul S., 80
Rial, Dr. William, 76
Riemann, Dr. Stanley P., 113
Rimmer, Robert H., 173
RNA, 35
  synthesis of, 25–27, 37–38, 146
  as transmitter of memory, 247–254
  to combat senility, 256
Robin, Dr. Eugene D., 81–82
Robinson, Dr. Bryan W., 200
Rock, Dr. John, 104
Rogers, Dr. Carl, 291
Rostand, Dr. Jean, 66–67, 108, 116, 119–120, 129–131, 133, 143, 146, 160, 256
Rowson, Dr. L. E., 121

Sade, Marquis de, 167
Sakel, Dr. Manfred, 227
Salin, Dr. Olle, 79
Salk, Dr. Jonas, 7, 11, 23, 29
Sanctity of Life and the Criminal Law, The, 72
Sapirstein, Milton R., 180
Sartre, Jean Paul, 175
Schafer, Curtiss, R., 207
Schmidt, Dr. Francis Otto, 256–257
Science and Human Values, 289
Selye, Dr. Hans, 39
Sem-Jacobsen, Dr. Carl W., 196, 204
sex
  capacity for
    affected by cerebral chemistry, 212–213
    affected by ESB, 198, 205
    men and women compared, 180–181
  clinical research on, 166–167, 170, 181
  contemporary attitudes, 165–175
  impotence, treatment of, 219
  and love, 183
  predetermination of, 115–116, 138
sex-linked diseases
  elimination of, 138–139
Sex, Science and Society, 115
Shapley, Dr. Harlow, 7, 13, 143
Shaw, George Bernard, 31, 145, 282
Sheen, Bishop Fulton J., 80
Sherman, Dr. Jerome K., 154
Sherrill, Rt. Rev. Henry Knox, 95
Sherrington, Sir Charles, 244
Shettles, Dr. Landrum B., 104, 113, 114
Shumway, Dr. Norman, 58, 84
sickle-cell anemia, 28, 158
Silverman, Dr. Frederic N., 176
Silvin, Francis, 154
Simon, Dr. William, 170, 172, 185
Singer, Dr. Marcus, 43, 68
Siodmak, Curt, 264
Situation Ethics, 79, 182
Skinner, Dr. B. F., 179, 221, 284
Smythies, Dr. John, 236
Snow, C. P., 289
Some Guiding Principles for Human Studies, 95
Soul, 71–74, 269
space travel
  artificial astronauts, 268
  cyborg astronauts, 275–276
  embryonic passengers, 124
  genetic adaptation for, 142–143
sperm banks, 154–157
sperm freezing, 154
Sperry, Dr. R. W., 246–247
Spiegelman, Dr. Sol, 37
Spurway, Dr. Helen, 111–113
Steig, William, 283
Steward, Dr. Frederick C., 128–129, 133
Story of O, The, 167
Stress of Life, The, 39
Stubbs, Major General Marshall, 217
Sturgis, Dr. S. H., 40
Suarez, Dr. Ramon, 82
Suda, Dr. I., 68
superovulation, 138–139
  in animal breeding, 121–123
  egg banks, 159
Sutton, Dr. Samuel, 248, 249
Swann's Way, 246
Sweet, Dr. William, 196, 201–202

326

Index